TEXTILE FABRIC FLAMMABILITY

TEXTILE FABRIC FLAMMABILITY

S. Backer, G. C. Tesoro, T. Y. Toong, and N. A. Moussa

The MIT Press

Cambridge, Massachusetts, and London, England

A report on research programs sponsored by the Government
Industry Research Committee on Flammable Fabrics and the Office
of Flammable Fabrics, National Bureau of Standards, Washington,
D.C.

PUBLISHER'S NOTE

This format is intended to reduce the cost of publishing certain
works in book form and to shorten the gap between editorial
preparation and final publication. The time and expense of
detailed editing and composition in print have been avoided by
photographing the text of this book directly from the authors'
typescript.

Printed in the United States of America.

Library of Congress Cataloging in Publication Data.
Main entry under title:

Textile fabric flammability

   Bibliography: p.
   Includes index.
   1. Inflammable textiles. I. Backer, Stanley,
1920-
TS1449.T366   677'.0287   75-23061
ISBN 0-262-02117-X

TABLE OF CONTENTS

FOREWORD   vii

ACKNOWLEDGMENTS   ix

NOMENCLATURE   xi

CHAPTER I   HAZARDS OF FLAMMABLE FABRICS   1
     1.1 Introduction   2
     1.2 Evaluation of Hazards   8
     1.3 The GIRCFF Program   13
     1.4 Accomplishments and Status   19

CHAPTER II   FABRIC PROPERTIES RELATED TO FLAMMABILITY   21
     2.1 Introduction   22
     2.2 Technological Descriptions of Fabrics   23
     2.3 Thermophysical Properties   31

CHAPTER III   IGNITION   107
     3.1 Introduction   108
     3.2 Heat Transfer Analyses   111
     3.3 Pre-Ignition Heat Transfer through a
          Fabric Skin Simulant System   124
     3.4 Radiative Ignition   131
     3.5 Flaming Ignition   152
     3.6 Comparison of Radiative and Flaming
          Ignition   174
     3.7 Summary of Results   182

CHAPTER IV   FLAME SPREADING   187
     4.1 Introduction   188
     4.2 Flame Propagation in Fabric Strips   195
     4.3 Flame Spreading in Garments   213
     4.4 Comparison of Flame Spread Results on
          Fabric Strips and on Garments   226
     4.5 Effects of Fabric Parameters (not
          considered by GIRCFF) on Flame Spreading   227
     4.6 Overview   232

CHAPTER V BURN INJURY   235

    5.1 Introduction   236

    5.2 Burn Injury Criteria   241

    5.3 Previous Studies on Burn Injury
         Potential of Fabrics   265

    5.4 Burn Injury Potential of GIRCFF Fabrics
         in Garment Form   273

    5.5 Burn Injury Potential of GIRCFF Fabrics
         in Strip Form   287

    5.6 Mechanisms of Heat Transfer to Skin   306

    5.7 Overview of Studies of Burn Injuries   326

CHAPTER VI SUMMARY, CONCLUSIONS, AND RECOMMENDATIONS   338

    6.1 Summary   339

    6.2 Conclusions   356

    6.3 Recommendations for Future Work   362

GIRCFF REPORT REFERENCES   365

REFERENCES (General)   370

LIST OF FIGURES   376

LIST OF TABLES   381

INDEX   384

FOREWORD

This report comprises an overview of the research conducted from 1971 to 1973 by four separate laboratory groups under the sponsorship of the Government Industry Research Committee on Flammable Fabrics (GIRCFF) and of the Office of Flammable Fabrics of the National Bureau of Standards. Its purpose is to present a comprehensive summary of their reports and to correlate wherever possible both their theories and experimental results.

No attempt is made to include all the details of the four grantee programs, yet this report is intended to serve as a "stand-alone" document which of itself meets the needs of textile technologists and fiber scientists concerned with flammability of apparel fabrics.

Readers who are concerned with further details of heat transfer analysis and experimental equipment featured in the various group studies will find it desirable to review the original reports on which this overview is based. These are referenced at the conclusion of this document. The laboratories in this cooperative program were: (1) Factory Mutual Research Corporation (FMRC); (2) Georgia Institute of Technology (GIT), School of Mechanical Engineering; (3) Gillette Research Institute (GRI), Harris Research Laboratories Department; (4) Massachusetts Institute of Technology (MIT), Fuels Research Laboratory, Department of Chemical Engineering.

This overview study has been conducted by a fifth, separate, group in the Department of Mechanical Engineering at M.I.T. It focuses on the specific aspects of the flammability problem as studied by the above grantee laboratories, namely, fabric ignition (Chapters 2 and 3), flame spreading (Chapter 4), and burn injury to the skin (Chapter 5).

The phenomenon of ignition is a complex one, involving such processes as heat transfer and thermal decomposition governed by fluid mechanics and chemical kinetics. For synthetic fabrics, thermoplastic behavior adds to the complexity. Treatment of the ignition phenomenon outlined in this study is essentially physical. It omits considerations of chemical kinetics and focuses on prediction of ignition time of fabrics under known conditions of thermal exposure.

Flame propagation can be thought of as a continuing ignition process and numerous factors influence burning speeds of textiles in fabrics and in garments. This study deals only with experimental aspects of the flame propagation phenomenon.

The third phase of this study, concerned with infliction of thermal burns on a human body, likewise involves heat transfer and thermal decomposition. In this phase, attempts are made to predict depth of burn damage on the basis of thermal history at the skin surface. The extensive experimental results reported throw light on the mode and extent of heat transfer from burning fabrics and garments to skin simulants.

A word concerning the formulation of this report. The results presented here are based entirely on the work of the four GIRCFF grantee laboratory groups noted above. The only additional experimental results furnished by the overview group are the scanning electron photomicrographs considered desirable to round out the characterization of the test materials of the program. The analytical developments reported are primarily those of the four laboratory groups, although certain portions of the theoretical discussion are drawn from the classical heat transfer literature. Interpretations of experiment and theory are either compiled from referenced GIRCFF reports, or reflect the independent views of the overview group. It is hoped that text format will make it clear as to whose views are being discussed at any given point.

Cambridge, Massachusetts   The Overview Group
January 15, 1975   S.B.   G.C.T.
                               T.Y.T.   N.A.M.

ACKNOWLEDGMENTS

Funds for this study were made available by the American
Textile Manufacturers Institute, the Man Made Fibers Producers
Association and the National Cotton Council and Cotton Incor-
porated. Government funds were provided by the National Science
Foundation through the National Bureau of Standards, Department
of Commerce. This financial support is gratefully acknowledged.

The authors wish to thank Mr. Henry Tovey, Chairman of the
Government-Industry Research Committee on Flammable Fabrics, for
his guidance, support, and encouragement during the course of
this work.

Appreciation is extended to Dr. Stelios Arghyros for
reviewing the manuscript and for preparing most of the scanning
electron photomicrographs presented in Chapter 2. Thanks are
due Dr. R. L. Woolard of the Fiber Surface Research Laboratory,
E. I. du Pont de Nemours & Company, Inc., for photomicrographs
of fabric #9.

Dr. John F. Burke of the Massachusetts General Hospital has
reviewed selected portions of this report, as has Professor
I. V. Yannas of M.I.T. Their suggestions and comments have
been most helpful.

Mr. Joseph Ting and Mr. Christopher Brogna have prepared
many of the drawings included herein. Mrs. Patricia Malhotra
and Miss Dorothy Eastman have effectively managed the editorial
activities of the project and have typed the bulk of the manuscript.

NOMENCLATURE

a	Stretch factor of depth-magnification of skin-simulant
c	Specific heat, $W \cdot s/g{}^\circ C$)
ΔE	Activation energy for damage rate process in skin, cal/mole
h	Convection heat transfer coefficient, $W/cm^2{}^\circ C$
H_b	Heat of burning in air, cal/g or cal/cm^2
H_{bt}	Heat transferred to skin, cal/g or cal/cm^2
H_c	Heat of combustion in 100% oxygen, cal/g or cal/cm^2
H_e	H_b/H_c
k	Thermal conductivity, $W/cm^2{}^\circ C$
k/δ	Thermal conductance, $W/cm^2{}^\circ C$
N_{Bi}	Biot number $\equiv h/(k/\delta)$
N_{Fo}	Fourier number $\equiv (k/\delta)\tau/(p\delta)c$
P	Frequency factor of damage rate process in skin, sec^{-1}
q_t, q_c, q_r, q_p	Total, conductive, radiative and condensation (of pyrolytic products) heat flux density, on skin surface, $cal/cm^2 s$
q^*_{rad}	Normalized radiative heat flux $\equiv \dfrac{\tilde{\alpha}\, W_o}{(k/\delta)(T_{i,m} - T_o)}$
q^*_c	Normalized convective heat flux $\equiv \dfrac{2h_c(T_f - \bar{T})}{(k/\delta)(T_{i,m} - T_o)}$
Q	Thermal dosage on skin surface, cal/cm^2
R	Universal gas constant = 1.987 cal/mole $\cdot$ K

T	Temperature, °C or °K
$\bar{T}$, T_{avg}	Average fabric temperatures over the ignition interval, °C
V	Flame speed, cm/sec
W_o	Radiative power flux, W/cm^2
x	Coordinate normal to fabric surface, cm
y, z	Integration variables.

Greek Symbols

$\tilde{\alpha}$	Absorptivity
δ	Fabric thickness, cm
ε	Emissivity
κ	Extinction coefficient, cm^{-1}
$\kappa\delta = \kappa*$	Optical thickness
$\tilde{\rho}$	Fabric density, g/cm^3
$\rho\delta$	Mass per unit area, g/cm^2
ρ	Reflectivity
σ	Stefan-Boltzmann constant = 5.71×10^{-5} $erg/cm^2 s °K^4$
$\tilde{\tau}$	Time, second
τ	Transmissivity
ψ	The angle subtended by a point in the fabric with the normal to the simulant surface
Ω	Thermal damage integral in epidermis and dermis.

Subscripts

c	Convection
e	Exposure
f	Flame
g	Gas
i	Ignition
m	Melting
p	Pilot-Ignition
rad	Radiative
s	Skin; Surface
t	Total
o	Initial

Abbreviations Used Synonymously

Cellulose Acetate/Nylon --- Acet./Nyl.; Ac./Nylon
Polyester --- PE; PET
Polyester/Cotton --- PE/C; PE/Cotton; PET/C; PET/Cotton
Polyester/Viscose Rayon --- PE/Ray.; PE/Rayon

Institutional Abbreviations

FMRC --- Factory Mutual Research Corporation
GIRCFF --- Government Industry Research Committee on Flammable
          Fabrics
GIT --- Georgia Institute of Technology
GRI --- Gillette Research Institute
MIT --- Massachusetts Institute of Technology

CHAPTER I

HAZARDS OF FLAMMABLE FABRICS

1.1 INTRODUCTION   2

1.2 EVALUATION OF HAZARDS   8

1.3 THE GIRCFF PROGRAM   13
     1.3.1 Study of Ignition   14
     1.3.2 Flame Propagation   15
     1.3.3 Burn Injury   17

1.4 ACCOMPLISHMENTS AND STATUS   19

CHAPTER I

HAZARDS OF FLAMMABLE FABRICS

1.1 INTRODUCTION

Current interest in consumer product safety encompasses clothing and textiles as well as a variety of other products, devices, appliances and equipment. Recent enactment of consumer protection legislation and the setting of federal standards for flammability of clothing and other textile products reflect an increasing awareness of the potential fire hazards of fibrous materials. But it cannot be said that this awareness is a phenomenon of the last year or two. [1]+

Actually, clothing and textile fires have been reported in history and literature, but not much has been written until the mid-20th Century about the measures available to reduce hazard from such fires. In the United States, the Cocoanut Grove holocaust alerted public officials to the need for establishing flammability specifications for materials used in public buildings. During World War II the Quartermaster Corps, facing the need for an immense quantity of "textile housing" (i.e. tents) specified fire resistance as a required characteristic in its FWWMR* finished cotton duck. Towards the end of World War II, the military were also concerned with fire retardant treatments for fabrics which would be suitable for garment applications, whether for special use conditions, or for protection from special exposure such as high energy radiation. Much of the research conducted or sponsored by the military at that time has been reported in the open literature[2]. However, hazard from flammable fabrics was not viewed as a general consumer problem until a number of tragic deaths from garment fires in 1952 stimulated Congressional activity, and led to the Flammable Fabrics Act of

---

+Numbers in brackets denote references in end of manuscript.

*Fire, water, weather, and mildew resistant.

1953. This Law remains a cornerstone in the legislative
network regulating the flammability of textile products
purchased directly by consumers.

The evolution of governmental regulations since the
Flammable Fabrics Act of 1953 was passed has been recently
reviewed by Ryan[3], who also discussed subsequent Federal
laws and statutes, and outlined the provisions of the Amended
Flammable Fabrics Act which was signed into law in December of
1967. This amended Act gave "the Secretary of Commerce the
authority and duty to set mandatory flammability standards as
needed to protect the public against unreasonable risk....
It authorized investigations of deaths and injuries, research,
studies of the feasibility of reducing flammability and the
development of test methods and devices". It related to
standards "applicable to wearing apparel and interior furnish-
ings for homes, offices and places of assembly or accommodation",
but not to vehicles of transportation.

The Flammable Fabrics Act of 1967 contained many unique
features. It provided that the Secretary of Commerce
promulgate fabric flammability standards which are "reasonable"
and "practicable". It specified a three-step process of
publishing notices and findings in the Federal Register prior
to promulgating final standards, and procedures for enforcement
after the standards are issued. The Law also provided that
the Secretary be assisted by a National Advisory Committee
comprised of consumer representatives, distributors, and
manufacturers.

As Tribus pointed out[4], "what distinguished this effort
in the area of fabric flammability is that the responsibilities
for setting standards are so diffuse. The keywords 'reasonable
protection' must have a definition which is acceptable to
producers, distributors, consumers, enforcement officials
and politicians, if we are all to be spared counter-productive
free-for-all battles.

3

"It is evident both from the testimony heard by Congress and the language of the law that absolute, complete protection without regard for cost cannot be a realistic goal. The Secretary of Commerce must, of necessity, be expected to decide on a reasonable trade-off between cost and degree of protection".

Some observers might have considered this to be a simultaneous pursuit of safety and industrial profit/goals, which in many instances, may lie in opposite directions. An alternative assessment is that Congress recognized that resistance to change is proportional to the rate of change being promoted, and that the forcing of a severe step-change in procedure could well result in a drastic impact on the industry structure, with consequent dislocations.

Government philosophy since that time however, has changed, as is evidenced in the passage of the Consumer Product Safety Act and in the establishment of the Consumer Product Safety Commission in late 1972. The act enpowers this government commission to issue standards protecting the public from unreasonable hazards and issuance of such standards can now take place without extensive prior interaction with industry.

Regardless of the regulation mechanics used for issuance of fabric flammability standards, it is essential that all proposed standards be based on comprehensive knowledge of the factors influencing fabric flammability. But, as was pointed out in a 1969 Conference on Flammable Fabrics[5], little or no basic information was available on this subject at the time.

Fortunately, the situation has improved during the past four years and much more is now known about textile flammability problems as a result of research programs on the subject. The purpose of this report is to discuss the theoretical background and the experimental evidence which have been developed as a result of specific programs monitored by the National Bureau of Standards of the Department of Commerce, through the Government Industry Research Committee on Flammable Fabrics [GIRCFF].

Although this document focuses on the work carried out
by four grantee laboratories working under the GIRCFF Program
[R1-R28], acknowledgement must be made of the immense effort
which has been concurrently devoted to the flammability problem
by individual industrial organizations working on their own
products, and by companies and institutions cooperating with
one another. The information Council on Fabric Flammability
has made a major contribution to the dissemination of information on the problem. Such professional societies as the
American Association of Textile Chemists and Colorists, and the
American Chemical Society have devoted major portions of their
technical meetings to the subject of fabric flammability. One
also notes that over 20 programs on fire research were being
sponsored during a recent year by the National Science
Foundation in its Office for Research Applied to National
Needs. Many of these grants were directly concerned with or
contributed to background knowledge on the subject of fabric
flammability. In addition, other government agencies have
active research programs in this field (e.g. the Department
of Defense, NASA, and the Department of Transportation).

The extent of industrial activity on the subject of
textile flammability can be detected by simply reviewing the
Annual Index of World Textile Abstracts [6] which, during 1972
shows more than 120 items on the subject of flame retardant
finishing and flammability and more than 190 items during 1973.
Many of the items on finishes cover industrial patents taken
out in the U.S. or in Great Britain. (Since the policy of
World Textile Abstracts precludes repetition of items published
in the open literature, or in the field of patents, the total
literature in this field far exceeds the estimate based upon
the WTA Index.) In addition, a Journal of Fire and Flammability
has been established in the U.S. Many recent books on the
problems of flammability include discussions of flame retardants

designed for fibers and textiles[7] and new flame resistant
fibers made from thermally stable polymers have been reviewed
in the proceedings of several symposia[8].

An expansion in literature is one immediate result of
increased industry and government effort on the subject of
flammable fabrics. The more important result has to do with
the establishment of new standards to protect the consumer of
textile materials, and particularly highly dependent users
such as young children. In the apparel area, the main thrust
of activity under the Flammable Fabrics Act has been towards
children's sleepwear garments, but new standards have also
been set with respect to flammability of carpets and rugs,
and mattresses. Standards for upholstered furniture, and
children's apparel generally are under consideration.

Meanwhile, the increased pace of research and development activity on problems of textile flammability has been
reflected in a number of technological developments: many
new flame resistant fabrics are being offered commercially,
or are on the threshhold of commercialization at this time.
Some of these developments are extensions or improvements of
earlier discoveries; some have originated from research on
materials for the space age; still others may represent a first
generation of experimental materials which will reach commercial
maturity in due course.

The development of inherently flame resistant fibers is
complex and costly for as Suchecki states[9] success with
respect to flammability behavior is a necessary but not a
sufficient condition for the commercial success of a new fiber.
It is clear that among the 50 properties that one might
associate with textile fibers, failure in any one of a dozen
critical properties is sufficient to exclude a fiber from
consideration as a commercial product. Nevertheless, flame
resistant fiber developments are at hand. Improvements have
been made in the well-known modacrylic fibers and several new

6

types have become available. Polyvinyl chloride fibers from
foreign sources--previously viewed essentially as curiosities
or specialty items in the U.S.--are being used for the manufacture of flame resistant fabrics. Flame resistant viscose
rayon fibers, acetate fibers and, more recently, polyester
fibers have become available in commercial or semi-commercial
quantities. The flame resistance of aromatic polyamides
(such as NOMEX) has been utilized in an increasing number of
applications, and advances have been made in the development
of polybenzimidazole (PBI) fibers. In addition, newly
discovered fibers from thermally stable polymers (e.g. Kynol,
PTO chelates) are currently being investigated.

Development of fire retardant fabric finishes has also
made significant progress. Several chemical systems for
finishing of 100% cotton are now utilized commercially, and
finished cotton is competing with fabrics made from flame
resistant fibers in the children's sleepwear market. Fabrics
made of 100% polyester and finished with flame retardant
chemicals are available. Among the products mentioned, some
will survive the tests of production costs and consumer
acceptance--others may fail. However, there is every
indication that additional new developments will be forth-
coming in the area of flame resistant textile materials, and
that industry's response to Federal legislation has shaped
new technology.

As Suchecki has written, "without any doubt, the course
of research and the conduct of the market place have dramati-
cally altered with a swiftness that has left many industry
people in a state of shock, and somewhat alarmed at the power
of the government to impose new standards and regulations". [9]

It is clear that opposing viewpoints of industry, the
regulating agencies and consumer advocates cannot be reconciled
a priori unless one views them and merges them in a decision-
making process based upon rational analysis. Tribus [4] has made

an attempt in this direction. His concept is that for a given
year, with a specified class of garments and a defined population of users, increased costs are generally associated with
increasingly stringent flammability standards. A graph for
such a trade-off can be easily conceptualized, and if compliance
with the standard is achieved, then the incremental cost of
attaining the level of such a standard can be compared with
the increase of safety obtained. The Tribus approach may be
used as a means of introducing the technical programs which
have been carried on under the sponsorship of the GIRCFF.

1.2 EVALUATION OF HAZARDS

Clearly, everyone who is concerned with problems of
flammable fabrics--the manufacturer, the retailer, the consumer,
the legislator, and the doctor--all would like to see the
hazard of burns from flammable clothing reduced. As Tribus[4]
has pointed out, the situation can be described as an exercise
in group risk-taking. The trade-off graph shown in Fig. 1-1

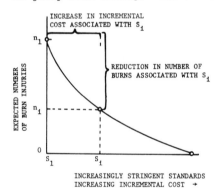

FIGURE 1-1 CONCEPTUALIZED TRADE-OFF ANALYSIS
(AFTER TRIBUS[4])

illustrates a reduction in the number of burns expected for a
given standard with an increase in the incremental cost
associated with that standard, assuming full consumer use of

the improved standard. (This graph implies that standards can be gradually made more severe and that there is a continuous relationship to hazards incurred in use of such fabrics.) For each reduction in number of injuries, we have an increase in incremental cost.

Figure 1-2 shows the decision tree discussed by Tribus to indicate that a decision-maker "can establish a value scale for measuring the outcome of his decision, and if he can assign the various probabilities with some degree of confidence, then the expected result of the decision can be evaluated".

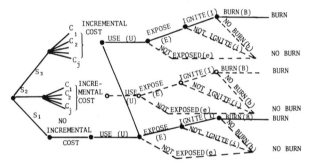

FIGURE 1-2   THE DECISION TREE (AFTER TRIBUS[4])

According to the decision tree of Fig. 1-2, Tribus has assumed three alternate standards of protection, S_1, S_2, and S_3. For each standard, one can hopefully calculate an incremental cost function. This function should include not only the additional direct cost of manufacturing the modified item, but also the effect of the new material on the dollar value in the eyes of the consumer, taking into account factors such as style, durability, washability, etc., which the consumer considers in making a purchase. Clearly, it is impractical to fix an overall single cost increment for any new product complying with a new standard (e.g. S_3) and so Tribus suggests the evaluation of added costs in terms of probabilities of incurring cost increases at levels C_1, C_2, C_3 to C_j for such a standard, S_3.

9

Once the direct dollar cost increase and the value reductions as viewed by the consumer are taken into account, there is a probability that the consumer will or will not use the product corresponding to the given standard (e.g. S_3) and at a given price. If the consumer does not use the product, he may, for example, turn to other material such as S_1 which does not conform to the standard, and use it for making articles of clothing. (Actually, where mandatory standards exist, it is found that alternative materials not meeting the standard are indirectly driven from the market place.)

If the consumer does use the product (e.g. a garment produced according to one of the more stringent standards), he will some time face exposure to heat sources that can cause ignition of that garment. The extent of exposure to ignition sources will depend on the life style of the individual, on his age, on the section of country in which he lives, on his economic status, and on a variety of other factors which are being defined on the basis of statistical information accumulated by NBS and by other agencies.

If a garment is exposed to an ignition source during its use, it may or may not ignite, and if it does ignite, it may cause burn injury to the wearer, it may self-extinguish, or it may be extinguished either by the wearer, or through the assistance of a second individual.

Tribus considers the probability that burn injury will occur upon adoption of a standard S_i and suggests use of the variation of this probability with the standard level, S_i, as a basis for the decision-making process for the setting of standards. As he points out, the total probability of starting with each more stringent standard S_i and moving through various paths (in Fig. 1-2) to occurrence of burn injury, equals the sum of probabilities associated with all such paths which lead to the burn.

In addition, depending upon what cost increment is involved, there is also a probability of injury resulting from use of non-improved material such as S_1 or, in other words, non-use of the

improved material S_3. The injury probability resulting from
non-use of S_3 must be added to the injury probability resulting
from actual use of S_3 to give the total injury probability
resulting from the setting of the standard S_3.

Since each path to burn injury consists of a joint
sequence of events, the path probability can be taken as the
product of probabilities of the several events which must occur
in sequence in order for the final event (burn injury) to occur
at all. Thus, according to Tribus, the probability of burn
injury through any one path following upon the setting of a
standard with expectations of <u>cost</u> increments, a corresponding
<u>use likelihood</u>, a <u>likelihood of exposure</u> to dangerous sources, <u>a</u>
<u>likelihood of ignition</u> from such exposure and a <u>likelihood of</u>
<u>burn injury</u>, may be written as the product of the five probabil-
ities.

One might argue that Tribus' concept of series of events
leading to burn should be expanded in the last term mentioned
above, i.e. for burn probability. Once ignition of a fabric
occurs, it may be desirable to consider the probability of flame
spread, taking into consideration the fabrics as combined in
garments, the garment design and fit, and also wearer reactions.
It is entirely possible that a flame, once ignited, will move
a short distance and then self-extinguish. Thus, there is a
probability of extinguishing spreading garment flames, given a
particular fabric combination, garment fit and design, and
certain wearer reactions. Actually, it would be appropriate to
consider the probability of <u>not extinguishing</u> rather than
extinguishing, so that the appropriate linkage in joint events
can be maintained, in a fashion similar to the other
probabilities. Finally, one may consider the probability of
thermal injury or burns resulting from the intensity of
garment combustion.

If we consider all these individual probabilities, including the suggested modification that the final probability of
burn be taken as the product of the probabilities of propagation,

11

of non-extinguishment, and of thermal injury resulting from spreading garment combustion, we can then determine the total probability of burn injury. Tribus suggests that once the various probabilities have been assigned, the expected number of burn cases may be computed by multiplying the probability of burn for a given standard by the number of individual users, perhaps dividing them into various age groups for which various probabilities might differ.

The major point in Tribus' presentation is that this technique represents a method of computing expected numbers of burn cases at various levels of expenditure for protection. Obviously, the desirable solution would be for the cost of flame resistant fabrics and garments to be made so low that the number of burn cases would be reduced to zero. Unfortunately, the development of such flame resistant products is a complex problem. Incorporation of flame retardant additives or monomers in the fiber-manufacturing process has a profound effect on fiber properties and on processing cost. Treatment of fabrics with flame retardant finishes also affects fabric cost, aesthetics and performance properties. If the maximum degree of safety were prescribed, the consumer would have to be satisfied with a very limited selection of products, and under these circumstances, for the case of voluntary standards, the probability would exist that usage of the given material at the higher cost would approach zero. The net result may be a greater number of burn injuries than for a lesser standard.

Tribus emphasizes that his analysis is used merely as an illustration, but it does have the virtue of bringing the individual components of the problem into relationship with one another and it permits subdivision of the problem into its constituent parts, each of which may be investigated either in the field, or in the laboratory. It suggests indices which go beyond a corpse count and which may be utilized to measure progress being made in reducing the number of burn injuries.

12

1.3 THE GIRCFF PROGRAM

The Government Industry Research Committee on Fabric
Flammability was established in 1970 at the initiative of
Dr. Tribus (then Assistant Secretary of Commerce for Science and
Technology). He recognized the need for merging the viewpoints
of government and industry in the effort to study the fundamental
phenomena underlying ignition and burning in diverse textile
products, and to approach objective definitions of hazard.
Tribus succeeded in obtaining funds for the program from govern-
ment (through the National Science Foundation) and from industry
(through the American Textile Manufacturers Institute, the Man
Made Fibers Producers' Association and the National Cotton
Council). Technical representatives of each of the supporting
organizations were appointed to form the nucleus of the Committee
(GIRCFF) which was chaired by a representative of the National
Bureau of Standards, and also included technical advisors from
NBS among its members. Research proposals were solicited in
response to a work statement prepared by this Committee, and about
50 proposals were received. These were reviewed critically and a
selection was made of those proposals which ultimately formed
the GIRCFF program discussed in this report. Simply stated,
the objective of the GIRCFF program has been an enhanced understanding of hazards from flammable fabrics. In the deliberations
of the Committee, the work statement was prepared, the research
proposals were reviewed and the experimental programs were
monitored with this major objective in mind.

The program of the GIRCFF has recognized from the beginning
the validity of the probabilistic approach to the problem of
fabric flammability and burn injury. The work plan of the
Committee has focused on two probabilistic elements in the series
of events which lead to human injury from burning clothing:
(a) the probability of ignition of a given fabric or garment,
given a specific thermal exposure and, (b) the probability of
burn injury to the wearer of the garment following fabric
ignition.

13

The probability of thermal exposure for typical activity in a given life style was not studied under the sponsorship of GIRCFF, for it was concluded that the provisions of the amended Flammable Fabrics Act did not consider means for controlling the probability of exposure. Since the Act did include means for controlling probability of ignition upon a given exposure and of burning following ignition by setting appropriate standards, it was logical for the program to focus on ignition, and on burn injury resulting from flame propagation in fabrics and garments. The great complexity of the problem inevitably led to investigations under conditions which were oversimplifications of real fire situations neglecting, for example, the effects of wearer reactions. Tests were run on fabric strips as an example of simple geometry amenable to theoretical analysis, and on more complex shapes such as garments which were a step closer to a realistic environment.

Some work was carried out to examine the possibility of wearer detection of the presence of heat sources through the fabric thickness, thus permitting evasive action prior to ignition. Pre-ignition heat transfer studies were carried out at the Factory Mutual Research Corporation Laboratories and these indicated that ignition from a radiative or a flaming source generally would occur before the wearer of a fabric is conscious of pain. It was thus concluded that further study of pre-ignition heat transfer would not advance a solution to the hazards of flammable fabrics.

1.3.1 Study of Ignition (Chapters 2 and 3)

According to the GIRCFF plan, the phenomenon of ignition was approached from different viewpoints by the laboratories working under the aegis of the GIRCFF. The Fire Hazard and Combustion Research Laboratory of the School of Mechanical Engineering, Georgia Institute of Technology (GIT) undertook experimental and analytical studies of the ignition times of 20 fabrics selected and furnished by the GIRCFF. Ignition

14

(or melting) times were measured under specified heat fluxes from radiative and flaming heat sources. To compare experimental and theoretical results, Georgia Institute of Technology (GIT) determined the physical and thermal properties of the 20 GIRCFF fabrics. This provided the most extensive collection of thermophysical data on textiles reported to date.

In their studies of the GIRCFF fabrics, Factory Mutual Research Corporation (FMRC) determined minimum ignition (or melting) times under conditions of flaming ignition. Furthermore, FMRC investigated the dependence of these minimum times on fabric variables (such as weight and porosity), flame variables (such as burner size, fuel rate, air-fuel ratio), presence of skin stimulant and sample orientation in the gravity field.

To model more realistic ignition hazards, the Gillette Research Institute (GRI) determined the ignition behavior of sleeve-like GIRCFF fabric specimens exposed to a prototype gas stove burner. The dependence of ignition times on fiber content, fabric structure, relative humidity and location above the stove was investigated.

Using the results obtained above, the Massachusetts Institute of Technology (Overview Team) proposed possible parameters to rank the GIRCFF fabrics with respect to the heat input needed for ignition or melting. These parameters depend on the thermal and physical properties of the fabrics and may provide guidelines for designing fabrics with longer ignition times and hence a lower probability of ignition.

1.3.2 Flame Propagation (Chapter 4)

Once ignition has occurred, flame propagation is the next link in the hazard chain and some effort was devoted in the GIRCFF program to the study of flame spread over selected fabrics and fabric combinations. Linear flame velocity was measured in laboratory specimen tests and clothed mannequins were used to

15

study the extent of fabric (and body) burn area in simulated garment fires.

The Fuels Research Laboratories at M.I.T. focused on tests using fabric strips, and their observations on flame velocities were but a part of a more extensive investigation of heat transfer from burning fabric specimens to a skin simulant, and of the skin burn damage to be expected from such exposure.

Attempts were made to conduct single layer burn tests for all samples in the GIRCFF series. Selected fabrics which did not burn freely in a single layer experiment (because of shrinking, melting and dripping), were layered together with a cotton knit fabric and flame velocities were recorded for these composite specimens. Because of the irregular burning characteristics of the thermoplastic fabrics of the GIRCFF series, the M.I.T. Fuels Laboratory group concentrated on the 100% cotton and the PET/cotton fabrics in the series. They studied the relationship between flame velocity and fabric weight, as well as other factors. Observations were also recorded as to the melting, dripping, and irregular burning which took place in experiments with composite samples.

The Gillette Research Institute studied fire spread for the GIRCFF fabrics in garment form. Their data on flame propagation were also but a part of a broader study of heat transfer and burn damage incurred during simulated garment fires. They observed the effect of garment fabric weight on the rate of flame spread. The influence of garment design and garment fit, (corresponding to fabric-skin simulant spacing in the M.I.T. laboratory fabric strip tests) was also observed.

The Gillette group reported failure to observe flame propagation in single garments made from light weight thermoplastic garments. In tests on composite systems including garments made

16

from thermoplastic and from cellulosic fabrics, the flame velocities observed generally fell below those recorded for 100% cellulosics of comparable weight. The order of layering the composites, i.e., whether the thermoplastic was used for the outer or for the inner garment, was also shown to have an effect on rate of flame spread.

1.3.3 Burn Injury (Chapter 5)

Both the M.I.T. Fuels Laboratory and the Gillette Research Institute studied heat transfer from burning GIRCFF fabrics to skin simulants, in one case using a fabric specimen, and in the other, using clothed mannequins. In addition, as a part of another program, the Gillette group has conducted burn tests involving anaesthetized rats, and has made histological studies of burn depths so as to correlate levels of heat dosage received by the mannequin with extent of burn.

One phase of the M.I.T. Laboratory program involved the development of special mechanical skin simulants which manifest heat absorption and transfer characteristics similar to those of the human body. A principal feature of these simulants is their capacity to "stretch" the thermal path so as to permit measurement and recording of the time-temperature history at various positions in the simulant which corresponded to actual, very shallow depths in the human skin. The 80-micron depth below the skin surface, which corresponds to the base of the epidermal layer, is of particular interest. It has been reported by other workers, that the temperature history of this critical depth can be used to characterize the onset of second-degree burns, and the medical profession emphasizes the critical influence of the area of second-degree burns on the chances of survival for individuals injured as a result of clothing fires.

Another skin simulant developed by the M.I.T. Fuels Laboratory provided surface sensors for measuring heat flux to the simulant plane, and these flux measurements were used to characterize heat transfer into theoretical models of the human skin, as well as to calculate temperature-time histories at various

17

depths in the hypothetical body exposed to the impinging flux.

The flux measurements were also used to describe the heat transfer history with movement of the flame across the sensor in the simulant. They were supplemented by careful temperature measurements in the fabric simulant air-gap, which were then used to subdivide the total flux into its conductive, radiative and condensative components. The relative importance of the various component mechanisms of heat transfer was determined as a function of fabric-skin spacing. The M.I.T. Fuels Laboratory group considered the usefulness of determining total heat dosage impinging on the skin as a basis for predicting or characterizing the depth of irreversible skin damage, which in turn is a rough measure of burn injury.

A similar conclusion was reached by the Gillette group on the basis of measurements of burn depth in anesthetized live rats held in an opening in an instrumented Transite board at a fixed distance from an upward burning fabric. Accordingly, the total thermal dosage measured by surface sensors was used by GRI as a basis for characterizing second-degree burn regions in clothed mannequin tests, and to evaluate burn hazard in terms of area burned (suffering second-degree damage, or worse) after specified exposure times.

The laboratory specimen tests at M.I.T. showed the total thermal skin damage caused in burning of cotton and PE/C fabrics to be proportional to fabric weight. This phenomenon was primarily dependent on the effect of fabric weight on flame velocity, not so much because of the occurrence of hotter, or larger flames in burning the heavier fabric, but rather because the duration of exposure to thermal dosage is longer for the heavier fabric where the flame movement over the sensor is slower.

Similar observations of higher heat dosages from burning of heavier fabrics were made in the Gillette studies of clothed mannequins. However, changes in garment size and design influenced the rate and extent of flame propagation, as well as the average heat transfer to cause second-degree burn areas. Thus, the heat transfer from burning garments was more difficult to analyze and correlate with material design features than in the case of fabric strip tests.

The clothed mannequin experiments were particularly infor-
mative in the case of fabric combinations where thermoplastic
or even flame-resistant fabrics were used for outer garments or
for undergarments, along with garments made from cellulosic
fabrics. Here it was shown that when supportive reinforcement was
provided during burning of thermoplastic fabrics, the heat dosage
was dependent on total fabric weight. But when a thermoplastic
- or a flame-retardant-treated material - was used for the undergarment, the level of heat transfer was considerably less.

These results were generally consistent with results of the
M.I.T. Fuel Laboratory tests on composite strip specimens - although
the latter tests were restricted to the case of a cotton underfabric with a range of materials used on top. This does not mean
that the absolute levels of heat dosage provided in mannequin
tests matched those of fixed-space laboratory-specimen burn-
experiments: The conditions of burning were too different to
expect such agreement, but the consistent results in the two
testing systems define some important trends.

1.4 ACCOMPLISHMENTS AND STATUS
The objective of the GIRCFF program broadly defined as an
enhanced understanding of hazards from flammable fabrics - has
been an ambitious and open-ended one. The work reviewed in this
report has made significant contributions to this enhanced
understanding and, in the context of other concurrent research
activities, provides a good basis for future progress in the
development of methods, materials, systems and standards designed
for improved fire safety in textile apparel.

Perhaps even more importantly, the GIRCFF experience has
demonstrated that technical representatives from Government
agencies and from Industrial organizations can productively merge
their efforts in formulating programs towards a well-defined
common goal, and further, can effectively coordinate research conducted in a variety of industrial, institutional and academic
laboratories, directed towards that goal. The mechanism of

19

GIRCFF may well serve as a model for joint government/industry endeavors towards the solution of technical problems of national significances.

Much work remains to be done on the problems of flammable fabrics: Some questions formulated at the onset of the GIRCFF program have still not been answered; some additional questions have been raised by program results; and others are evolving from changes in the spectrum of commercially available materials, from experience with newly set standards, and from new statistical information on burn injuries. Recommendations for future work outlined in Chapter 6 of this report are based on consideration of some of these questions. It is not now known whether the work needed can continue to be jointly funded by government and industry, nor whether the GIRCFF experience can be continued in some form to design new cooperative ventures. It is clear, however, that optimum solutions of the technical and legislative problems pertaining to flammable fabrics depend on a continuation of the spirit, if not the letter, of the GIRCFF approach.

CHAPTER II

FABRIC PROPERTIES RELATED TO FLAMMABILITY

2.1 INTRODUCTION   22

2.2 TECHNOLOGICAL DESCRIPTIONS OF FABRICS   23

2.3 THERMOPHYSICAL PROPERTIES   31

    2.3.1 Fabric Mass per Unit Area   32
    2.3.2 Specific Heat of Fabrics   32
    2.3.3 Thermal Conductance   35
    2.3.4 Ignition and Melting Temperatures   40
    2.3.5 Fabric Radiative Properties   44
    2.3.6 Heat of Burning   47
    2.3.7 Effect of Temperature on Fabric Structure   51
    2.3.8 Thermophysical Properties - Summary   56

CHAPTER II

FABRIC PROPERTIES RELATED TO FLAMMABILITY

2.1 INTRODUCTION

As with any solid, a textile fabric exposed to a heat
source will experience a temperature rise under influence of
the resultant heat transfer. If the temperature of the source
(either radiative or gas flame) is high enough and the net
rate of heat transfer to the fabric is great, pyrolytic decomposition of the fiber substrate will soon occur. The products
of this decomposition will include combustible gases, non-
combustible gases and carbonaceous char. The combustible gases
mix with the ambient oxygen. This mixture ignites, yielding a
flame, when its composition and temperautre are favorable.
Part of the heat generated within the flame is transferred to
the fabric to sustain the burning process and part is lost to the
surroundings.

It is well established that burning entails physical
processes involving transient heat transfer (with numerous
complex combinations of radiation, convection and conduction)
as well as chemical processes of thermal degradation, and of
oxidation of combustible products of pyrolysis. The major
emphasis in the GIRCFF studies of ignition and burning of
commercial textile fabrics has been on the physical side, with
considerable effort directed toward understanding the nature
and effects of rapid heat transfer from heat source to fabric
and from burning fabric to skin or skin simulant. The analysis
and experiments relating to these two circumstances of heat
transfer are discussed in separate chapters of this report in
the context of selected, commercially available materials.

This chapter focuses on the properties of the GIRCFF
fabrics which have been introduced into the analysis and which
have been measured by several of the grantee laboratories.

Technological properties are provided to permit identification
of material composition and of geometric, i.e. textile, configuration, but the main emphasis in this handbook-like
chapter is on the thermophysical properties which influence
heat transfer to a fabric from an external source, the
propagation of flame, and the extent of resultant injury to
the wearer of clothing. These thermophysical properties
include thermal conductance, specific heat, area density,
ignition and melting temperatures, infrared reflectance and
transmittance, and heat released during the combustion process.
Technological description and properties of the fabrics include
fiber composition, interlace structure (weave or knit formation),
surface finish (e.g. napped or brushed), chemical finish, color,
commercial designation of fabric, yarns per inch and porosity
(i.e. air permeability). Finally, since it was recognized
that many of the above properties are dependent on ambient
conditions during measurement (specifically on temperature and,
to some extent, on relative humidity) efforts were made to
measure certain critical properties under a range of conditions corresponding to extremes incurred during usage or
during the actual transient heat transfer.

2.2 TECHNOLOGICAL DESCRIPTIONS OF FABRICS

Textile fabrics vary so much in basic interlacing structure as well as in fiber and yarn arrangement at their surfaces,
that it is essential to formulate a logical, simple classification scheme as a guide to judging the nature of materials
selected for experiments and analysis. Many such classification schemes have been developed in textbooks on textile
structures and in dictionaries and thesauri having to do with
textile materials. For example, Hearle [10] classifies fabrics
and other laminae on the basis of their structural components.
He lists five such components, (a) interlaced yarns, (b) non-
interlaced yarns, (c) fiber assemblies, (d) fibrous sheets, and

(e) non-fibrous sheets. The most common textile structures
are composed of yarn systems interlaced together either in
weaves or in knitted structures. Textile structures made
of bonded yarn sheets are produced as commercial items, but
they are not very common. Wool felts, needled felts and
bonded fiber fabrics represent examples of fiber assemblies
without yarn components. Leather and paper are classified
as fibrous sheets, while non-fibrous sheets include such
materials as plastic films, rubber sheets, metal foils, etc.
By combining these five components, one can identify a wide
range of textiles and textile-like materials as to structure.

Dictionaries and thesauri are likewise good sources for
structural classifications of textile material. One such
thesaurus [11] provides systematic classification of fibers,
yarn systems and fabric systems. It differentiates between
multifilament yarns and spun yarns, between continuous filament
(flat yarns) and textured filament yarns. Further, it differentiates between cotton spun, woolen spun, and worsted spun
yarns, all of which vary in surface characteristics. Fabrics
are classified by weave type or by knitting structure. Of
particular interest in the study of flammability is the designation of flat woven fabrics, of ordinary weft knitted or
warp knitted fabric, and of pile fabrics constructed on a woven
base, a knitted base, a stitch-bonded base, or tufted base.

Classification of textile materials for the specific purpose of studies in flammability behavior has been suggested
by Bernskiold [12]. His classification scheme is shown
in Fig. 2-1, and differentiates between the structure of the
base cloth and the surface of the fabric system. For example,
the base fabric may be composed of a yarn system, or it may be
a fiber assembly which does not possess yarn components. The
type of fiber assembly of which the base cloth is formed and the
composition of the surface structure, whether comprised of
unorganized fiber pile or of organized yarn pile are then

24

SINGLE-LAYER MATERIAL		Surface consisting of			
		Yarn system	Fiber system	Fiber pile	Yarn pile
Base consisting of	Yarn system	Knitted fabric Woven fabric	Needle-punched carpet with fabric base Raised fabric	Cut-pile fabrics Cut-pile carpets	Loop-terry fabric Loop-pile carpet
	Fiber system		Needle-punched fabric Nonwoven fabric Wadding	Cut-pile carpet with spun-bonded fabric	Loop-pile carpet with spun-bonded fabric

FIGURE 2-1   SINGLE-LAYER MATERIALS BUILT-UP FROM SUB-LAYERS OF DIFFERENT STRUCTURES [12].

characterized. Figure 2-1 indicates the variety and combinations of yarn and fiber systems which are used in the base fabric and surface layers of common textile materials. This variety is of considerable importance to the subject of flammability in relation to such characteristics as true area of surface exposure, cloth porosity, dimensional stability, thickness, etc.

The problem of clothing fires is obviously complicated by the fact that garments are rarely worn in single layers and so one must provide a designation for fabric combinations in different end uses, as shown, for example, in Fig. 2-2[12]. The layering illustrated characterizes the common placement of various textile materials, and the black triangle indicates the application point of an igniting flame. For example, the stocking is shown in direct contact with the skin with little or no air layer between. A curtain near a window has a considerable air layer between it and the glass. An outer shirt and tight fitting underwear will have a layer of air between the two fabrics, but little spacing between the under-wear fabric and the skin, etc.

When the GIRCFF program was initiated, it was decided that experimental work should be carried out by all contractors on the same fabrics, selected so as to represent a broad spectrum of commercially important products, as is outlined in Table 2-1.

It is unfortunate, but not surprising, that only one flame-resistant fabric was included in the group (FR-treated cotton flannel). Few flame resistant fabrics were commercially available at the time and, furthermore, it was expected that important differences would be detected in the ignition behavior and in the potential contribution to hazard for widely used commercial fabrics varying widely in fiber composition, weight per unit area and porosity. An attempt was made to include fabrics of varying weight, napped and "pile" fabrics (flannel and terrycloth), and representative constructions made from

26

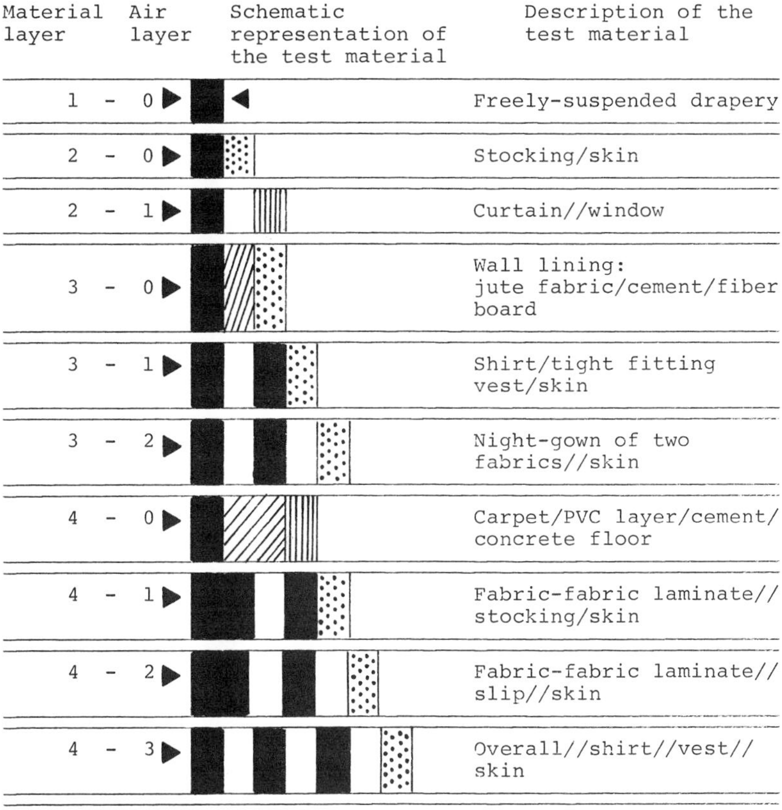

Material layer	Air layer	Schematic representation of the test material	Description of the test material
1 - 0 ▶		■ ◀	Freely-suspended drapery
2 - 0 ▶		▓▓	Stocking/skin
2 - 1 ▶		■ ▥	Curtain//window
3 - 0 ▶		▨▓	Wall lining: jute fabric/cement/fiber board
3 - 1 ▶		■ ■ ▒	Shirt/tight fitting vest/skin
3 - 2 ▶		■ ■ ▒	Night-gown of two fabrics//skin
4 - 0 ▶		■ ▨▥	Carpet/PVC layer/cement/ concrete floor
4 - 1 ▶		■ ■ ▒	Fabric-fabric laminate// stocking/skin
4 - 2 ▶		■ ■ ▒	Fabric-fabric laminate// slip/skin
4 - 3 ▶		■ ■ ■ ▒	Overall//shirt//vest// skin

FIGURE 2-2   SCHEMATIC REPRESENTATION OF THE TEST MATERIALS
CONSISTING OF 1-4 MATERIAL LAYERS.   ▶ IGNITING
FLAME, /NO AIR LAYER, //AIR LAYER, [12].

TABLE 2-1

COMMERCIAL DESIGNATIONS OF FABRICS USED IN GIRFF FLAMMABILITY RESEARCH

(as of November 3, 1970)

Weight Oz/sq. yd.	Cotton	PE/Cotton	PE/Rayon	Acetate	Ac./Nylon	Acrylic	Nylon	PE	Wool
2-5	batiste flannel: untreated FR treated T-shirt	batiste 65/35 T-shirt 65/35 shirting 50/50		tricot	tricot	jersey	tricot text. woven blouse taffeta		
5-8	terry	slacks 65/35 untreated; DP treated	slacks 65/35 DP treated					double knit	flannel
8+	denim								

natural fibers, from thermoplastic fibers, and from blends of
the two. While this range of constructions covered only 4 out
of the 12 structural examples listed in Fig. 2-1, it included
the most commonly used apparel fabric structures and thus
served as a reasonable experimental base. It was felt that
other structural categories could be dealt with at a later
date if it were shown that the geometry of such textile materials or their method of manufacture (e.g. use of binders in
non-wovens) had significant influence on ignition and burning
behavior.

The expectation that the variables included would cover a
broad enough range of "hazard potential" has been only partially
realized. It is now apparent that greater insight would have
been gained from studies including fabrics of much lower flammability (e.g. Nomex, Kynol, modacrylics), and it is clearly
indicated that such materials should be added to the list of
fabrics for any future work contemplated.

The fabrics used by GIRCFF contractors[*], listed in Table
2-2, were obtained from manufacturers' production and subsequently scoured to remove soluble residues and contaminants.
No attempt was made to determine "purity" analytically. In
those instances where durable chemical finishes were present
(e.g. fabrics #19, #1 and #15), their composition was not
disclosed or analyzed. Undyed fabrics were obtained whenever
possible to avoid the potential complications arising from
differences in dyestuff content.

Effects of foreign material on flammability were not
identified in the program and constitute a separate, although
important, phase of the overall problem. This might be investigated in future work with the objective of defining effects of
dyes, finishes, or materials deposited in laundering.

Table 2-2 lists details of technological description
and of selected fabric properties. The methods for determining

---

*Fabrics 24† and 25† were not in the original GIRCFF series
but were added in the GRI study.

TABLE 2-2 FABRICS USED IN GIRCFF RESEARCH

Mfg. Class.	GIRCFF No.	Fiber Comp.	Color	Mechan. Finish [a]	Chem. Fin. [b]	Weave [c]	Wt. [d]	No. Yns. [e]	LOl [f]	Perm. [g]
Batiste	10	100% Cotton	Purple	---	---	Plain	2.04	187		
Batiste	10	100% Cotton	Yellow	---	---	Plain	1.90	192	.162	3.4
Batiste	10	100% Cotton	Red	---	---	Plain	1.98	189		
Batiste	10	100% Cotton	Blue	---	---	Plain	1.98	196		
Flannel	18	100% Cotton	White	Napped	---	Plain	3.69	88	.171	1.19
Flannel	19	100% Cotton	White	Napped	Fire Rtd.	Plain	4.21	89	.252	0.09
T-Shirt, Jersey	5	100% Cotton	White	---	---	Knit	3.97	75	.171	1.18
Terry Cloth	9	100% Cotton	White	---	---	Pile	7.42	87	.158	1.20
Denim	4	100% Cotton	Navy	---	---	Twill	8.66	109	.169	0.07
Batiste	17	65/35% Poly/Cot	White	---	---	Plain	2.53	173	.179	1.89
Shirting	16	50/50% Poly/Cot	White	---	---	Plain	3.91	102	.182	0.92
T-Shirt, Jersey	8	65/35% Poly/Cot	White	---	---	Knit	4.82	77	.167	1.03
Untreated Slack	6	65/35% Poly/Cot	White	---	---	Twill	6.99	139	.171	0.18
Durable Press Slack	1	65/35% Poly/Cot	White	---	DP Treated	Twill	7.07	141	.171	0.18
Durable Press Slack	15	65/35% Poly/Ray	Brown	---	DP Treated	Plain	6.62	91	.179	1.71
Tex. Woven Blouse	2	100% Polyester	Yellow	---	---	Twill	2.19	260	.190	0.70
Double Knit	3	100% Polyester	White	---	---	Double-Knit	5.83	121	.200	5.1
Taffeta	14	100% Nylon	White	---	---	Plain	1.67	184	.247	0.58
Tricot	12	100% Nylon	White	---	---	Knit	2.47	98	.198	7.4
Tricot	11	80/20% Acet/Nylon	White	Brushed	---	Knit	2.67	100	.179	0.40
Tricot	13	100% Acetate	White	---	---	Knit	2.44	80	.179	6.8
Jersey Tube Knit	7	100% Acrylic	Gold	---	---	Knit	2.64	52	.162	2.6
Flannel	20	100% Wool	Navy	Napped	---	Twill	6.48	57	.261	0.92
Pique	24+	100% Cotton	White		Fire Resist.	Waffle	4.00	---	---	---
Plain	25+	100% Nomex	Green		---	Plain	4.10	---	---	---

[a] Information from OFF, NBS

[b] From Manufacturer.

[c] Information from OFF, NBS.

[d] Information from OFF, NBS - Oz./sq.yd.

[e] Information from OFF, NBS - Yarns/sq. in.

[f] Information from Burlington Industries, Inc.

[g] From Factory Mutual Research Corporation - (ft/sec)(lb/ft^2).

+ GRI code

30

these properties are discussed later in this chapter and the
properties will be listed again for different ordering of
the fabrics: such repetition might help the reader in dealing
with the collections of data presented in the report.

2.3 THERMOPHYSICAL PROPERTIES

Numerous experimental determinations of thermophysical
properties were carried out on the GIRCFF fabrics. For the
most part these measurements were conducted by the research
staff of the Georgia Institute of Technology (GIT). This
chapter discusses these measurements and provides experimental
results relating to the GIRCFF fabrics. It is, therefore, primarily a compilation of data reported by GIT. In some cases,
however, measurements were made by other laboratories as a
part of the overall program and this will be noted where
appropriate.

The GIT group considered ignition to occur in a textile
material when either its surface temperature, or its mean
temperature reached a certain critical value. In the study
of ignition of thin-fabrics exposed to radiation heat sources,
surface temperature was used as the ignition criteria. Thus,
as a first approach, the phenomenon of igniting fabrics under
radiation exposure was considered as a process of heating a
thin slab of inert material to a point where its surface
reached a critical ignition temperature. Once this ignition
temperature is established, one can attempt to determine how
much heat must be transferred to the fabric to raise its
temperature from room temperature to this critical ignition
temperature. This heat input equals the product of the
required temperature rise, the mass of the fabric, and its
specific heat. The time of ignition for a fabric exposed to a
given set of heat transfer circumstances, can then be determined simply by dividing the rate of heat transfer into the
total heat required to meet the ignition condition.

31

This oversimplified picture of the initial stages of the ignition process suggests the nature of properties to be measured for prediction of ignition time in fabric systems. These properties include fabric mass per unit area, specific heat, thermal conductance, and reflectance and transmittance. In addition, as the GIT group points out, it was necessary to measure convective heat transfer film coefficients for the fabric system as applicable to forced convection during gas flame ignition, and to free convective losses during radiative heating.

2.3.1 Fabric Mass Per Unit Area

GIRCFF fabric weights (oz/sq.yd.) have been listed in Table 2-2 on the basis of data furnished by the Office of Flammable Fabrics (NBS). Measurements were also conducted by the GIT group and their mass per unit area data based on ten weighings of 5" by 5" desicated samples, are presented in Table 2-3. Whereas the data of Table 2-2 are grouped according to fiber type, and each group is listed in order of increasing weight, the data of Table 2-3 are listed in order of the arbitrary GIRCFF fabric coding, i.e. 1 to 20. There is considerable advantage to the form of listing of Table 2-2, but since most of the data reported by the four grantee laboratories follow the numerical order of the GIRCFF codes, this overview report will do likewise.

Table 2-2 shows four cotton batiste specimens, each of different color, slightly different weight and yarn count. Table 2-3 indicates that the GIT group worked specifically with the purple batiste #10.

2.3.2 Specific Heat of Fabrics

The GIT group derived the specific heat of the GIRCFF fabrics as a function of temperature on the basis of enthalpy

32

TABLE 2-3 FABRIC IDENTIFICATIONS AND SPECIFIC MASS [R8]

GIRCFF No.	Classification	Fiber Composition	Color	Finish	Specific Mass mg/cm^2
1	Durable Press Slack	65/35% PE/C	White	DP treat.	23.49
2	Text. Woven Blouse	100% Poly.	Yellow	-	7.51
3	Double Knit	100% Poly.	White	-	20.91
4	Denim	100% Cotton	Navy bl.	-	29.63
5	T-Shirt, Jers.	100% Cotton	White	-	13.71
6	Untreated Slack	65/35% PE/C	White	-	23.57
7	Jers. Tube Knit	100% Acry.	Gold	-	15.13
8	T-Shirt, Jers.	65/35% PE/C	White	-	16.19
9	Terry Cloth	100% Cotton	White	-	26.48
10	Batiste	100% Cotton	Purple	-	6.65
11	Tricot	80/20% Acet. Nylon	White	-	11.31
12	Tricot	100% Nylon	White	-	8.91
13	Tricot	100% Acet.	White	-	9.40
14	Taffeta	100% Nylon	White	-	5.66
15	Durable Press Slack	65/35% PE/ Rayon	Brown	DP treat.	22.82
16	Shirting	50/50% PE/C	White	-	13.14
17	Batiste	65/35% PE/C	White	-	8.55
18	Flannel	100% Cotton	White	-	12.88
19	Flannel	100% Cotton	White	Fire ret.	14.89
20	Flannel	100% Wool	Navy bl.	-	19.93

measurements at various fabric temperatures. They graphically smoothed the curves of enthalpy vs. temperature data and represented them by means of a power polynomial. The results were then differentiated with respect to temperature to obtain specific heats at various temperature levels.

To obtain enthalpy data, the fabrics were heated in a modified Setchkin furnace[13]. This furnace was originally developed for measurement of the ignition characteristics of plastics and was thermostatically controlled with controlled air circulation. The commercial version of the furnace* was modified by GIT as indicated below. Longer legs were fitted beneath the furnace and an easily removable insulated bottom was added. Provision was made for a steel tube in lieu of a ceramic liner tube. Aluminum discs were used to line the new removable bottom so as to provide a complete lining of metal within the furnace. A stainless steel wire mesh basket, 3" in diameter, and 4.2" long, was used to contain the sample in the furnace. The same basket was used to hold the sample as it was taken from the furnace and quickly deposited in a calorimeter placed beneath the furnace. Temperatures of the furnace and of the fabric samples were monitored by thermocouples.

The calorimeter was a 2150 cc silvered Dewar flask. The sample receiver of the flask was 3.125" in diameter and 5" long, made of copper. An overlapping copper lid was provided to cover the sample receiver and was submerged partially when the receiver was lowered into the water bath of the calorimeter. A solid foam lid was also provided to cover the flask. A platinum resistance thermometer was supported from the cover of the flask.

In the test procedure, about 80 gms of the sample fabric were rolled up tightly into a cylindrical shape and placed in the stainless steel basket of the Setchkin furnace. Its weight was determined before and after the calorimetric measurements. A thermocouple placed in the center of the fabric sample was

*Customs Scientific Instruments

34

used to determine when the fabric came into thermal equilibrium with the furnace temperature. It took 4 hours, or more, to reach this equilibrium. The furnace bottom was then removed and the sample basket dropped into the copper sample receiver in the preconditioned calorimeter. The receiver was then covered by the copper lid and the calorimeter was closed. It took from 45 to 90 minutes for the fabric sample to come to equilibrium temperature in the calorimeter and the calorimetric temperatures were recorded during this period with the bath being manually agitated to equalize temperatures.

The procedures for conducting these experiments and for treating the resultant data are discussed in detail in Appendix B2 of [R4]. Detailed graphs of specific enthalpy vs. temperature for ten of the GIRCFF fabrics are also furnished in the body of this reference. Specific heats for the 20 GIRCFF fabrics are shown in Table 2-4, and it is seen that they vary between 0.91 and 2.68 Ws/g°C measured over a temperature range between 50°C and 250°C.

The GIT group collected specific-heat data from the litera- ture for ten different fiber materials, each listed generically without reference to fabric structure or experimental techniques. It was found that the range of values reported was in general agreement with measurements on the GIRCFF fabrics. (The data of other workers are presented in Appendix C1 of [R4] and are based on references [14, 15, 16, 17]). Only one reference, however, provided values at varying temperatures.

2.3.3 Thermal Conductance

Thermal conductance of the GIRCFF fabrics was measured on a modified guarded hot plate in which a known heat flux was passed through a sample of known thickness and the conductance was determined from the temperature drop across the specimen. The fabric sample was mounted between two 5" by 5" brass equalizing plates which had been ground and lapped flat. A

TABLE 2-4 [R8]

SPECIFIC HEAT OF FABRICS IN W s/(g C) AS FUNCTION OF TEMPERATURE

GIRCFF No.	Temperature °C							
	50	75	100	125	150	175	200	250
1	1.19	1.24	1.29	1.34	1.38	1.43	1.47	1.53
2	1.42	1.42	1.42	1.42	1.42	1.42	1.42	-
3	1.03	1.10	1.17	1.25	1.36	1.50	1.67	2.16
4	1.17	1.23	1.29	1.35	1.40	1.46	1.50	1.55
5	1.05	1.19	1.33	1.47	1.62	1.77	1.91	-
6	1.10	1.16	1.21	1.26	1.32	1.39	1.48	-
7	1.27	1.40	1.53	1.71	1.92	2.23	-	-
8	1.23	1.37	1.44	1.54	1.65	1.75	1.86	-
9	1.13	1.21	1.32	1.45	1.53	1.61	1.64	-
10	1.37	1.50	1.64	1.77	1.91	-	-	-
11	1.45	1.45	1.45	1.45	1.45	1.45	1.45	-
12	1.51	1.70	1.90	2.09	2.29	2.42	2.68	-
13	1.24	1.38	1.51	1.65	1.78	1.92	2.05	-
14	1.59	1.66	1.73	1.69	1.86	1.96	2.13	-
15	0.91	1.03	1.15	1.26	1.37	1.47	1.55	1.66
16	1.16	1.21	1.26	1.31	1.37	1.41	1.44	-
17	1.27	1.33	1.40	1.47	1.53	1.60	1.66	-
18	1.44	1.50	1.56	1.63	1.69	1.75	1.82	-
19	1.33	1.42	1.51	1.60	1.69	1.78	1.87	-
20	1.27	1.27	1.27	1.27	1.27	1.27	1.27	-

heater was mounted below the second equalizing plate and
a Transite insulating plate sandwiched between two more
equalizing plates was placed below the heater. Finally, a
guard heater was provided below the second set of equalizing
plates to eliminate a temperature differential across the
transite plate and insure that the heat supplied by the
middle heater flowed up through the fiber sample. Radial
guard heaters were also provided to minimize edge losses.

The upper plate resting on the fabric exerted a known
pressure on it and by adding weights to the upper plate it
was possible to establish conditions of varying pressure on
the sample during conductance measurement. The temperature
in each of the equalizing plates was measured by 5 thermocouples potted in slots in the plate and the thermocouples were
monitored by a microvolt potentiometer. The heat flux through
the sample material was determined from power measurement for
the sample heater. Full details of the guarded plate apparatus
and the test procedures are contained in Appendix B3 [R4].

Thermal conductance was measured on single layers of the
fabric over a temperature range of about 50 to 220°C and at
contact pressures ranging from 528 N/m^2 to 866 N/m^2. Thermal
conductance is plotted as a function of temperature in Fig.
2-3 and it would appear that for all 19 of the GIRCFF fabrics
tested, conductance increases as the temperature of the test
is increased. All of the data in Fig. 2-3 were obtained under
contact pressure conditions of 528 N/m^2. Similar data for thermal conductance of 10 selected fabrics in the GIRCFF series
are listed in Table 2-5. These latter data were obtained under
conditions of contact pressure of 866 N/m^2 and it is seen that
conductance is generally higher at higher contact pressures, as
has been noted many times in the literature. The only exception
to this trend is the case of sample #20, the wool fabric. Notice
that Table 2-5 includes a comparison of mass/unit area data
obtained at the National Bureau of Standards with those obtained

37

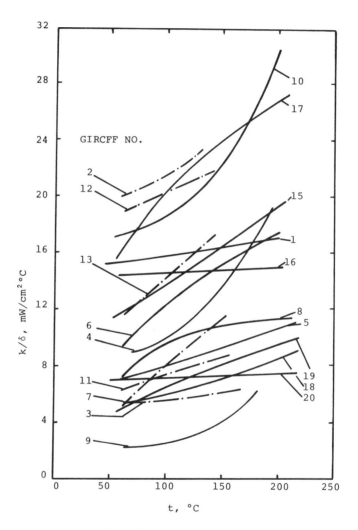

FIGURE 2-3. THERMAL CONDUCTANCE AS FUNCTIONS
OF TEMPERATURE [R8]

38

TABLE 2-5

THERMAL CONDUCTANCE SECONDARY GIRCFF FABRICS

AT CONTACT PRESSURE OF 866 N/M^2 [R8]

GIRCFF Fabric Number	Mass per Unit Area mg/cm^2		Original Fabric Thickness cm	Mean Temperature °C	Specific Conductance mW cm^{-2}C^{-1}
	NBS	GT			
1	24.04	23.49	0.0464	88.7	15.71
3	19.82	20.91	0.0953	92.0	5.46
4	29.44	29.63	0.0575	90.2	13.05
6	23.77	23.57	0.0476	88.9	14.53
7	8.98	15.13	0.0791	88.3	7.93
9	25.23	26.48	0.2080	95.3	2.98
14	5.68	5.66	0.0110	88.2	27.80
15	22.51	22.82	0.0421	89.0	15.37
16	13.29	13.14	0.0321	90.7	19.72
20	22.03	19.93	0.0721	91.7	7.00

at GIT. Measurements of the original fabric thickness expressed
in centimeters are also included in Table 2-5. A summation of
thermal conductance results for the remaining 10 GIRCFF fabrics
is given in Table 2-6 showing the effect of both temperature and
pressure.

The GIT group has provided an extensive literature survey
on measurements of thermal conductance of textile materials,
indicating that considerable research has been done on this
property since the early 19th Century. As GIT points out, heat
transmission through fabrics is a complicated process involving
conduction through the fiber, conduction and some convection
through interstitial air spaces, radiation, and even conduction
through absorbed moisture. Thus, thermal conductivity as applied
to a structure such as a textile fabric has no precise meaning.
In the literature numerous investigators have considered the
following parameters to affect thermal conductivity of fabrics:
fiber composition, density, mechanical pressure, moisture
content, temperature, treatment of the fiber and fabric, history
of usage of the fabric, and presence of multiple layers. The
GIT group has discussed the influence of these variables in some
detail in Appendix C2 [R4].

2.3.4 Ignition and Melting Temperatures

Ignition and melting temperatures were measured at GIT by
placing the fabric in a modified Setchkin Furnace flushed by
preheated air. The self-ignition temperature was taken to be
the lowest initial temperature of air passing around the specimen
which could be ignited without the need for an external pilot
flame. The pilot ignition temperature represented the lowest
temperature of air passing around the specimen which could be
ignited by a small external pilot flame.

Since there has been some discussion of the meaning and
validity of the quantity entitled "ignition temperature", it
is appropriate to quote the discussion of the GIT group concerning its measurement. "Self ignition occurs at the onset of

40

TABLE 2-6   THERMAL CONDUCTANCE OF PRIMARY GIRCFF FABRICS [R4]

GIRCFF Fabric Number	Original Fabric Thickness cm	Mass Per Unit Area g/cm^2 x 10^3 N.B.S.	G.T.	Pressure 1 528 N/m^2 k/δ mW $cm^{-2}C^{-1}$	$\overline{T}$ °C	Pressure 2 866 N/m^2 k/δ mW $cm^{-2}C^{-1}$	$\overline{T}$ °C	Pressure 3 1370 N/m^2 k/δ mW $cm^{-2}C^{-1}$	$\overline{T}$ °C
2	0.0201	7.41	7.51	20.26 21.02 22.79	73.0 93.1 124.6	29.43 21.03 22.74	62.8 83.3 86.3	22.51	75.8
5	0.0529	13.43	13.71	7.32 8.81 10.82	70.9 130.5 206.0	7.31 11.28	57.4 79.8	13.02	76.5
8	0.062	16.31	16.19	7.63 10.50 11.13	68.4 129.6 205.9	7.99 9.13	56.0 81.7	10.74	75.4
10	0.0177	6.91	6.65	17.48 20.19 29.67	67.8 128.4 195.8			25.36	76.5
11	0.0614	9.05	11.31	6.60 7.08 8.08	75.2 95.2 130.9	7.46	86.2	8.13	77.7
12	0.0271	8.36	8.91	19.30 17.00 20.93	76.1 96.3 122.6	17.35 20.80	56.1 92.2	17.89	73.5
13	0.0304	8.26	9.40	12.21 13.93 15.86	72.1 94.9 125.1	18.82	91.7	17.56	76.9
17	0.0193	8.56	8.55	17.34 22.48 26.75	72.6 128.0 199.3			30.41	77.3
18	0.0713	12.49	12.88	5.55 6.72 8.74	73.0 134.4 205.3			7.06	78.6
19	0.0692	14.25	14.81	5.38 7.55 9.46	73.0 134.1 203.5			8.55	77.8

41

spontaneous exothermic reactions between the fabric decomposition products and the oxygen which is present in the boundary
layer surrounding the fabric sample. The oxidation gases have
reached the activation energy corresponding to the combustible
gas composition. The energy imparted on the gases is the excess
energy supplied from the furnace via the air surrounding the
sample over the energy absorbed by the sample, either as
sensible heat, or during endothermic decomposition prior to
ignition. Consequently, the self-ignition temperature depends
not only on the air flow rate which affects both the gas
composition (oxygen index) and the boundary layer temperature
profile, but also on the heating rate which dictates the time
lag between furnace and fabric temperatures, and the degree of
decomposition at the time of ignition."

"An increase in ignition temperature with increasing air
flow is the result of more intensive mixing which yields rather
dilute concentrations at first during the heating process until
the time rate of decomposition increases with rising fabric
temperature. A decrease in ignition temperature with increasing
air flow would indicate oxygen deficiency at low air velocities."

"Low heating rates for the furnace should result in low
self-ignition temperatures as the difference between fabric
and air temperatures vanishes with decreasing heating rates.
The lowest possible air flow rate of 1.6 m/s was employed during
this program. Finally, the flash ignition temperature is the
lowest possible ignition temperatures due to the additional
energy supplied to the combustion gases by the ignition source",
Appendix B4[R4].

For thermoplastic samples, the GIT group recorded the
Setchkin furnace temperatures at which melting was first observed.
They did not measure self- or pilot-ignitions on the molten
polymer collected below the melting fabric in the pan-shaped
sample holder. A summary of the auto- and pilot-ignition and
of the melting temperature obtained on the GIRCFF fabrics is
presented in Table 2-7. Here, it is seen that pilot-ignition

TABLE 2-7 AUTO AND PILOT IGNITION AND MELTING TEMPERATURES [R8]

Temperatures in Centigrade

GIRCFF No.	Classification	Fiber Composition	Self-Ignition Melting	Pilot-Ign.
1	Durable Press Slack	65/35% PE/C	416-409	337-331
2	Textured Woven Blouse	100% Polyester	262-255 M**	
3	Double Knit	100% Polyester	255-246 M	
4	Denim	100% Cotton	297-287 290-280*	290-280
5	T-Shirt, Jersey	100% Cotton	311-201	311-301
6	Untreat.Slack	65/35% PE/C	416-409	345-337
7	Jersey Tube Knit	100% Acrylic	246-236 M	
8	T-Shirt, Jersey	65/35% PE/C	450-439 443-416	316-304
9	Terry Cloth	100% Cotton	308-297	297-290
10	Batiste	100% Cotton	439-429 443-434	351-339
11	Tricot	80/20% Acet./Nyl.	237-218 M	
12	Tricot	100% Nylon	255-237 M	
13	Tricot	100% Acetate	237-218 M	
14	Taffeta	100% Nylon	246-236 M	
15	Durable Press Slack	65/35% PE/Ray.	426-419	337-331
16	Shirting	50/50% PE/C	473-466	360-353
17	Batiste	65/35% PE/C	480-463	384-368
18	Flannel	100% Cotton	311-301	294-278
19	Flannel	100% Cotton	497-480	329-316
20	Flannel	100% Wool	480-463	329-316

*Check on reproducibility
**M represents melting

43

temperatures consistently fall below the self-ignition temperatures for the same fabric. The fabrics which melted but did not ignite in the Setchkin Furnace are marked with an M.

It is worth noting that fabrics of the same fiber composition, but different structure, do not ignite at the same temperature either in self-ignition or in pilot-ignition. Compare for example the cotton denim fabric #4 vs. the cotton knit fabric #5, or the cotton terry fabric #9 vs. the cotton batiste fabric #10. The difference in structure of these various cotton fabrics clearly influences self-ignition and pilot-ignition temperatures as measured at GIT.

The same observation holds for melting temperature data on thermoplastic fabrics of the same fiber composition. Melting temperature differences for the thermoplastic fabrics are much smaller than differences in ignition temperatures for non-thermoplastic cellulose-containing fabrics. Nonetheless, the GIT results obtained in the Setchkin Furnace appear to reflect fabric structural effects as variables of the system.

2.3.5 Fabric Radiative Properties

The GIT group measured reflectance and extinction coefficients for the various GIRCFF fabrics as properties which characterize interaction with incident radiative heat flux. In order to utilize these fabrics properly, it was necessary to determine the directional and the spectral description of the incident radiative heat flux in actual ignition tests. The tungsten filament temperature of the heating element used was approximately 3200° Kelvin and the spectral distribution was such that about 12% of the energy fell between the ultraviolet and the visible regions, 65% between the visible spectrum and the two micron wavelength and the remainder in the far infrared spectrum. This was the distribution reported by GIT for full power production of the heating element. At half-power production, these fractions varied slightly.

Fabric reflectance is dependent upon fabric geometry,
fiber material, treatment and, also, on the source characteristics.
The fraction of the incident flux reflected by the fabric is the
total hemispherical reflectance, while the portion of the
incident flux which enters the fabric is partly absorbed and
partly transmitted by the fabric. The transmitted energy can be
measured to evaluate the extinction coefficient of the material.

GIT measured the total hemispherical reflectance and the
extinction coefficient of the GIRCFF fabrics in a 6" integrating
sphere reflectometer. A Tungsten projector lamp provided
radiative energy which could simulate the spectral characteristics of the radiant heat source used in ignition testing. The
radiative energy entered the spherical reflectometer impinging
on the fabric sample at a given angle. A detector in the wall
of the enclosure picked up the magnitude of the energy reflected
indirectly from the reflector wall, including multiple reflec-
tions, and provided a signal proportional to the sample
reflectance. The reflectance measurement obtained on the basis
of the signal from a given fabric sample was compared with that
obtained from a reference sample of magnesium oxide. Absolute
reflectance without the use of reference was also determined.
Transmittance of the fabric was obtained "by varying the background behind the fabric, to either permit the transmitted
energy to be reflected by the enclosure, or to be absorbed by
a black opaque diaphragm". The GIT group used filters and
filament temperature adjustment to vary the spectral distribution of the incident radiation in the range of the heater
variation to be used in subsequent ignition tests (Appendix B5
[R4]).

Two principal sources were used: (1) the Tungsten filament
source at a temperature of 3160° Kelvin and (2) the infrared
source between 0.6 and 2.5 microns wavelength. Reflectance,
transmittance, and absorptance values for the 20 GIRCFF fabrics
are shown in Table 2-8 for samples in their original state.
Samples of the GIRCFF fabrics were then charred at the mean
temperature between room temperature and their respective
destruction temperatures. (Destruction temperatures are the

45

TABLE 2-8   OPTICAL PROPERTIES OF ORIGINAL FABRICS [R8]

Subscript 1 refers to Tungsten Filament Source, T=3, 160°K
Subscript 2 refers to Infrared Source, 0.6<λ<2.5μm

GIRCFF No.	Reflectance		Transmittance		Absorptance		Optical Thickness	
	$\tilde{\rho}_1$	$\tilde{\rho}_2$	$\tilde{\tau}_1$	$\tilde{\tau}_2$	α_1	α_2	κ^*_1	κ^*_2
1	0.605	0.522	0.211	0.289	0.184	0.189	0.375	0.296
2	0.560	0.501	0.276	0.346	0.164	0.153	0.272	0.208
3	0.560	0.619	0.250	0.238	0.190	0.143	0.335	0.276
4	0.450	0.491	0.105	0.137	0.445	0.372	1.106	0.855
5	0.533	0.521	0.288	0.296	0.179	0.183	0.284	0.282
6	0.581	0.581	0.220	0.284	0.199	0.135	0.388	0.223
7	0.536	0.549	0.262	0.159	0.202	0.292	0.340	0.855
8	0.562	0.560	0.264	0.276	0.174	0.164	0.298	0.272
9	0.681	0.623	0.206	0.231	0.113	0.146	0.274	0.288
10	0.418	0.406	0.361	0.394	0.221	0.200	0.280	0.236
11	0.523	0.508	0.320	0.329	0.157	0.163	0.230	0.232
12	0.447	0.397	0.380	0.433	0.173	0.170	0.214	0.188
13	0.490	0.443	0.344	0.390	0.166	0.167	0.227	0.203
14	0.348	0.348	0.508	0.434	0.144	0.218	0.136	0.234
15	0.515	0.413	0.259	0.331	0.226	0.256	0.375	0.340
16	0.613	0.499	0.219	0.294	0.168	0.207	0.338	0.312
17	0.485	0.464	0.364	0.372	0.151	0.164	0.198	0.208
18	0.599	0.573	0.231	0.251	0.170	0.176	0.326	0.312
19	0.590	0.602	0.229	0.197	0.181	0.201	0.348	0.428
20	0.336	0.365	0.119	0.102	0.545	0.533	1.149	1.226

self-ignition temperatures or, in the case of the melting
fabrics, the melting temperatures.) This procedure was
followed since it was considered that the optical properties
obtained from fabrics charred in this manner represent more
nearly the average behavior of the fabrics during the ignition
preliminaries. Table 2-9 lists the reflectance, transmittance,
and absorptance of the GIRCFF fabrics prepared in this manner.
It also lists the charring temperatures used for each of the
20 fabrics.

2.3.6 Heat of Burning

Eighteen out of the 20 fabrics selected for the GIRCFF
research had been studied by Yeh, using an isoperibol calorimeter [18]. Yeh's calorimeter had the dimension of 7.5" in
length and 3" in diameter, constructed with thin oxygen-free
copper sheet. A 2-junction thermocouple was mounted on the
calorimeter for temperature measurement and the EMF output
therefrom was fed to a 2-pen recorder equipped with an
integrator. The calorimeter was operated with a constant air
flow of 60 meters per minute at normal atmospheric pressure.
The fabric sample was cut into a piece 5" x 2" and mounted on
a copper holder held at a 45° angle during the measurement.
For fabrics which melt during burning, a fine metal gauze was
attached at the bottom of the holder to catch the melt. After
burning residue had been collected and weighed, data were
obtained for the heat released by the fabric and for the
percentage residue. These are furnished in Table 2-10.

The data show that the heat released during burning of
the three different colored cotton batiste fabrics (#10) was
essentially the same, indicating that dye-stuff differences
may have little effect on burning characteristics. There is
a slight difference, however, in the amount of residue measured
for the three different colors. Since undyed cotton fabrics
normally provide no residue in burning (as indicated by fabrics

47

TABLE 2-9   OPTICAL PROPERTIES OF CHARRED FABRICS [R]

Fabrics Irradiated by Tungsten Filament Source at 3,160°K

FABRIC GIRCFF No.	Reflectance ρ	Transmittance τ	Absorptance $\alpha = 1-\tau-\rho$	Optical Thickness κ^*	Char Temper. °C
1	0.366	0.285	0.349	0.490	219
2	0.582	0.243	0.175	0.305	134
3	0.506	0.294	0.200	0.306	134
4	0.401	0.088	0.511	1.305	162
5	0.491	0.284	0.225	0.348	162
6	0.339	0.288	0.373	0.510	219
7	0.113	0.167	0.720	1.115	230
8	o.346	0.308	0.346	0.460	230
9	0.627	0.131	0.242	0.660	162
10	0.219	0.394	0.387	0.413	230
11	0.566	0.256	0.178	0.303	134
12	0.465	0.346	0.189	0.289	134
13	0.492	0.326	0.182	0.291	134
14	0.399	0.408	0.193	0.222	134
15	0.239	0.158	0.603	1.040	247
16	0.322	0.437	0.241	0.256	247
17	0.255	0.320	0.425	0.520	252
18	0.527	0.232	0.241	0.431	162
19	0.248	0.254	0.498	0.688	252
20	0.105	0.286	0.609	0.725	252

*This source produces 11.3% visible and ultraviolet and 88.7% infrared radiation.

TABLE 2-10   HEAT RELEASE VALUES OF GIRCFF FABRICS*

GIRCFF No.	Classification	Fiber Composition	Residue,%	Oz/sq.yd.	Heat Release cal/g
10	Batiste	100% Cotton			
(purple)			3.3	2.04	3100
(yellow)			4.8	1.90	3140
(blue)			2.0	1.98	3130
			3.5(Ave.)	(Ave.)	3120
18	Flannel	100% Cotton	---	3.69	3270
19	Flannel	100% Cotton	45.0	4.21	1330
5	T-Shirt, Jersey	100% Cotton	---	3.97	3160
9	Terry Cloth	100% Cotton	---	7.42	3170
4	Denim	100% Cotton	---	8.66	3200
17	Batiste	65/35% PE/C	4.3	2.53	2970
16	Shirting	50/50% PE/C	5.5	3.91	2780
8	T-Shirt, Jersey	65/35% PE/C	2.9	4.82	3160
6	Untreated Slack	65/35% PE/C	9.1	6.99	2790
1	Durable Press Slack	65/35% PE/C	7.9	7.07	2660
15	Durable Press Slack	65/35% PE/Ray.	9.8	6.62	2400
11[1]	Tricot	80/20% Acet./ Nylon	9.5	2.67	3800
13[1]	Tricot	100% Acetate	10.5	2.44	3620
7[1]	Jersey Tube Knit	100% Acrylic	13.3	2.64	4990
20	Flannel	100% Wool	37.6	6.48	2210

[1]Wire gauze underlay

*After Yeh [18]

49

#18, #5 and #9), Yeh suggests that the presence of the dye-stuff may have some quenching effect on afterglow. But from the point of view of actual heat release, generally the cotton fabrics release an equivalent amount of heat regardless of weave structure or fabric weight.

The flame retardant finish on fabric #19, a cotton flannel, reduced heat released by a factor of about 50%, as compared to a similar untreated fabric (#18). The high residue yield for fabric #19 also indicates the effect of the flame retardant finish. Since the fabric burned in the calorimeter, releasing 1330 calories/gram, it was not considered by Yeh to be adequately treated.

Fabrics #17, #16, #8, #6, #1 and #15 represented blend materials, the first five being polyester/cotton and the last one, #15, a polyester/rayon blend. Of the five polyester/cottons, #8 which seemingly released the highest amount of heat, was the only knit material and had the second highest permeability in the group. More important, the knit structure, as can be seen in scanning electron micrograph[*], has lower twist and bulkier yarn structure than the remaining woven blends: The photo-micrographs of the woven fabrics show frequent intersection bindings along each one of the yarns, while such restrictive binding does not occur in the knit fabric, #8. Further, it can be expected that high temperatures will cause the polyester fibers in the blend to shrink and imbed themselves at the core of the yarn, exposing the cotton portion of the yarn blend in a bulked form. Such bulking can take place more easily in a knit structure.

Yeh points out that the efficiency of combustion of polyester in a blend is reciprocally dependent on the rate of combustion of that blend. Knit fabrics would tend to combust more slowly than flat woven fabrics, increasing the combustion efficiency of the polyester and thus the heat release for the knit fabric.

---

[*] Shown later in this chapter.

The polyester-rayon blend fabric gave lower heat release than the polyester-cotton blends. This was attributed to the fact that rayon has a lower fuel value than cotton.

In the case of thermoplastic fabrics, the melting and dripping caused considerable variation in the results, both with regard to heat release and residue yield.

2.3.7 Effect of Temperature on Fabric Structure

This section discusses briefly some of the changes in fabric form and dimension which occur when ambient temperature is increased. It is well known that thermoplastic fibers will shrink when heated, and that this shrinkage will increase linearly with the temperature of exposure. However, there is a temperature below which the fiber is relatively stable. This temperature of stability is generally dependent upon the heat setting condition to which the thermoplastic fiber was subjected before the shrinkage tests. The commercial heat setting of fibers and fabrics is designed to establish the stability temperature at a level above the conditions to which the fiber would be subjected in usage--either in washing or drying, or in ironing. The theory and practice of fiber and fabric setting is of considerable interest to the textile industry and this topic is receiving increasing attention in the textile literature [19].

Another variable influencing the shrinkage incurred by textile fibers is the restraining tension on the fiber during the shrinkage; if the restraining tension is negligible, free shrinkage of the fiber will proceed in a linear relationship to the temperature. However, if tensions exceed the shrinkage tensions, the heated fiber will extend, or creep, rather than shrink. On the other hand, if the fiber is extended slightly or allowed to contract slightly, and then subjected to increasing temperatures at a fixed length, it will undergo changes in tension. This type of behavior is pictured by Valk [20] in Figs. 2-4 and 2-5.

51

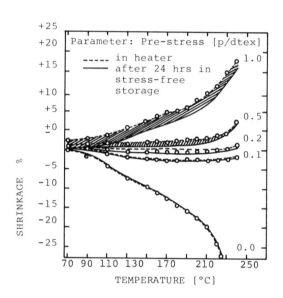

FIGURE 2-4   TEMPERATURE DEPENDENCE OF SHRINKAGE OF POLYESTER
YARNS (DIOLEN 100 DTEX 36 FILAMENTS) UNDER VARIOUS
PRE-STRESS CONDITIONS [20]

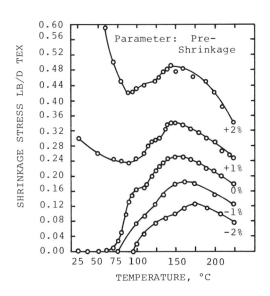

FIGURE 2-5   SHRINKAGE STRESS AS A FUNCTION OF TEMPERATURE
MEASURED FOR POLYESTER YARNS (DIOLEN 100 DTEX 36 FILAMENTS)
AT GIVEN PRIOR ELONGATIONS [20]

53

What does this mean in terms of fabric behavior during an exposure to ignition sources? It signifies that upon heating of the fabric, its fiber components will shrink in accordance with the restraining tensions imposed by the structure of the fabric. If the fabric is a very loose and open structure, the fiber will shrink under a constraint equivalent to about 2-3 mgs/den. And the shrinkage tendency of the fiber (under that fixed load) would be proportional to the increase in temperature above the stability temperature. In most instances, the fabric would tighten up before the fiber is able to shrink the full amount dictated by the temperature and with the tightening and locking of the fabric geometry, the fiber will convert from the behavior following one of the curves in Fig. 2-4 to the tension vs. temperature behavior shown in Fig. 2-5. For example, if the fabric had undergone shrinkage of 2% as a result of the early part of the heating exposure and the fabric tightened at the 2% shrinkage mark, then the tension in the fiber would rise starting at about 95°C and increase to a maximum at about 175°c. Increases in temperature beyond that would result in a reduction in fiber tension.

The peak tensions recorded in Fig. 2-5 reflect the initiation of crystal melting and the subsequent disintegration of the fiber and its mechanical coherence. It should be pointed out, however, that the curves of Fig. 2-5 reflect the original feed yarn of which the fabric is made. And if the fibers are subjected to a heat setting condition of about 200-220°C as is done in many setting processes, the thermal tension vs. temperature curve will tend to flatten considerably and the stress buildup will be accordingly suppressed.

When mixtures of thermoplastic and cellulosic fibers are used, it may be expected that thermal treatments will result in shrinkage of the thermoplastic fiber under fabric constraint due to locking of the weave, but also due to resistance to shrinkage afforded by the cellulosic fiber. The yarn system made of a combination of thermoplastic and cellulosic fibers

54

should therefore behave at high temperatures in a manner
similar to the behavior of a high-shrinkage, low-shrinkage blend.
It is well established in the technology that such blends of
high and low shrinkage fibers tend to bulk upon activation of
the shrinkage of one of the components. The high shrinkage
component goes to the center of the yarn, while the low
shrinkage component is bulked outward at the periphery of the yarn.

Still another form of dimensional and form instability is
observed in many textile structures. Conventional fabrics are
generally comprised of twisted yarn components and these yarn
components are inherently unstable under certain conditions.
For example, if one applies an axial tensile load to a twisted
yarn which is in a completely relaxed form, the yarn will tend
to develop an axial torque as well. This can be observed when
one hangs a dead weight on a twisted yarn and finds that the
weight begins to spin. An analysis of the yarn torques which
will be set up when one subjects a twisted yarn structure to
a tensile load has been made and reported upon in the literature[21]. And it can be expected that when fibers in the
twisted yarn are subjected to ambient conditions which activate
shrinkage forces (either through heat, or through moisture),
the yarn structure containing said fibers will experience not
only an axial tension and shrinkage tendency, but also an axial
torque and untwisting tendency. Torques can also be generated
in twisted yarns when high temperature exposures unlock
potential fiber torsional and bending moments resulting from
manufacturing processes. Shrinkage of yarns is a complicated
function of twist[22].

If such twisted yarns are present in a woven fabric
structure with a free edge or corner, the combined untwisting
torques which occur under shrinkage conditions in warp and in
filling will result in a curling moment being developed in the
fabric. Since, by definition, a free corner has no restraint
upon it, such curling torques will result in a rolling-up of
the fabric corners. This is a common phenomenon observed when
one takes a small square of fabric and moistens it, then hangs
it freely from one corner.

A similar kind of behavior occurs in knit fabrics, but
here the curling which takes place is more representative of
the effect of shrinkage tension in the fibers on unbending of
the yarns in the individual loops. Actually, curling of this
sort will occur quite freely in any knitted structures, even
without the presence of shrinking conditions to act on the fibers.
And this suggests that in many knit structures there is a considerable amount of bending energy present at all times. An
analysis has been conducted on the curling tendencies of woven
fabrics and this has led to the design of numerous balanced
structures so as to alleviate the curling tendencies, particularly
in long narrow strips such as medical bandages [23].

2.3.8   Thermophysical Properties - Summary

In the following section we restructure the data on thermophysical properties according to individual GIRCFF fabrics.
In addition to the table of properties provided for each fabric
(based primarily on GIT data) we include a scanning electron
micrograph. These micrographs show the original appearance
of the fabric, and in some cases the micrographs show fabrics
after heating to high temperatures.

The tabular data provide a description of the fabric and
its fiber composition, a listing of mass/unit area, and a
summary of specific heat values taken at several temperature
levels. In addition, thermal conductance of each material is
shown at various temperatures and at various contact pressures.
Also reported are the self-ignition and pilot-ignition temperatures or the melting temperatures. Finally, the absorptance,
reflectance, transmittance, and optical thickness are reported
for the original fabric condition as well as for the charred
condition.

The charring temperatures of the fabrics are taken as the
mean temperature between room temperature and the so-called
destruction temperature. The destruction temperature, obviously,
is much lower for fabrics which melt but do not ignite. Thus,

56

char temperatures for thermoplastic fibers are frequently of the order of 100°C lower than for many of the cotton or wool fabrics. The one exception to this is the acrylic fabric, #7, which is subjected to a charring temperature of 230°C. Also, it should be noted that several of the cotton fabrics have been exposed to charring temperatures at the relatively low level of 162°C. This includes samples #4, #5, #9 and #18, which represent some of the heaviest cotton fabrics of the test series. The point to be made here is that the optical properties of the charred fabrics are not, in a sense, isolated absolute properties of the GIRCFF fabrics, but rather they are optical properties of materials subjected to temperature conditions which have been selected, in turn, by the levels of temperatures determined in the self-ignition or melting tests.

A few words about the scanning electron microscope pictures. Sample #1 is a twill weave made of polyester-cotton blend yarns which have been treated with a durable press finish. It has 141 yarns/sq. inch and weighs 7.07 oz/sq. yd. This is one of the heaviest fabrics of the GIRCFF series and is one of the least permeable. Fabric #1 is similar in structure and appearance to another polyester-cotton blend twill, #6, which is virtually the same in weight, threads/inch, permeability and oxygen index. Fabric #1 is a durable press fabric; fabric #6 is untreated.

Sample #2, the textured woven polyester blouse material, which is a 2-up 1-down twill, possesses the largest number of yarns per square inch, 260. It is just a little over 2 ounces per square yard. As is evident in the photograph of the original state, this fabric is one of the tightest of the GIRCFF series, and its air permeability is at the low end of the scale. Nonetheless, one can observe the space between fibers in the vertical warp yarn, evidencing the effect of texturizing of the yarn, probably by the false twist process. As fabric #2 is subjected to a temperature of 200°C the crimp and waviness of the

57

individual fibers in the yarn bundle seem to increase and the
fabric appears to tighten up. When the fabric is exposed at
260° temperature, the filament crimp is almost completely
eliminated and the fabric tightens still further. In still
another photograph taken of the fabric after exposure to
230°C, one notes evidence of fiber melting and bridging between fibers.

Fabric #3 is a double knit polyester fabric made of textured yarn. It is a fairly heavy fabric, almost 6 oz/yd^2,
but it possesses less than half the number of yarns per square
inch, as compared to fabric #2. However, its air permeability
is much higher than that of fabric #2, as is evidenced in the
photograph. Also quite clear in the photograph is the crimped
nature of the individual filaments and the relatively high
surface exposure of the yarn bundle and knit loop structure.
Fabric #4, in contrast, is a heavy cotton twill fabric, 8.66
oz/yd^2. This fabric has the lowest air permeability of the
entire test series, undoubtedly due to the packing of the
highly twisted warp yarns (horizontal in the photograph). It
is worth while to contrast the relative exposure of the fibers
which comprise the open yarns of fabric #3 as compared to the
firm yarn structure of fabric #4. Also, fabric #4, in contrast
to fabric #3, shows the presence of a small number of fuzz fibers
at the surface of the yarn due to the staple nature of the yarn
construction while there are no free ends in fabric #3.

Fabric #5 is a single jersey knit structure used for T-
shirts -- evidenced by the considerable fuzz at the surface of
the fabric. Comparing the single knit of fabric #5 to the double
knit of fabric #3, one is given the impression that the single
knit fabric #5 provides much better cover, in its single layer,
than do the double layers of the double knit fabric #3. This
is evidenced in the much higher air permeability of fabric #3,
as compared with the T-shirt fabric #5. Note that fabric #5
with an air permeability of 1.18 is far more permeable than the
typical tightly woven fabrics, e.g. cotton fabric #4, the

relatively heavy denim twill or fabric #6, another heavy twill weave used for slack material.

Fabric #6 is a 7 oz/yd^2 polyester cotton blend, constructed in a twill weave. The staple nature of its yarn is evidenced in surface fuzz. However, the contrast between the yarn structure of fabric #5 and fabric #6 illustrates the effect on staple yarn structure of the twist level used in woven fabrics vs. those used in knit fabrics. Generally speaking, the yarns used in knit fabrics have a lower twist level, as is evident in the examples of fabric #5 vs. fabric #6. Also to be noted is the relative openness of fabric #5 as is verified by a ten-fold higher air permeability for fabric #5 vs. fabric #6.

Fabric #7 is an acrylic jersey knit (single knit) fabric which has a brushed surface treatment, and it may be compared with fabric #8, which is a polyester-cotton blend, made in a jersey knit structure for T-shirt use. And it, too, has a brushed surface treatment. The acrylic fabric #7 has 52 yarn/in^2 while the polyester-cotton jersey #8 has 77 yarn/in^2. The polyester-cotton blend appears to be a more tightly knit structure than the acrylic. The photographs imply a considerable difference in the air permeability between these fabrics and, as it turns out, the acrylic fabric #7 has more than twice the air permeability of the polyester-cotton blend. Again, the staple nature of the yarns of these fabrics is illustrated in the photomicrographs.

Fabric #9 is an all cotton cloth, unique amongst the GIRCFF series for its pile yarn loops on both back and front. The terry is a woven fabric of intermediate tightness, hence with intermediate air-permeability as compared with other woven fabrics. Because of its double pile structure it is the thickest of the test series, but not quite the heaviest. The cotton denim #4 is over 1 oz/yd^2 heavier. What is characteristic of the terry, as seen in the photographs, is the high surface exposure of the pile yarns--built into the cloth design to maximize water absorption. The structural difference also influences its thermal behavior, as evidenced by the tendency of fabric #9 to lie off the various curves reported in this study.

Fabric #10 is a light weight plain weave cotton batiste. It has the second highest number of yarns/sq.in. after the textured woven polyester fabric #2. Based on its high yarn count it might be considered one of the most tightly woven fabrics of the series. But a view of the scanning electron photomicrograph of fabric #10 in its original state indicates that this fabric actually has a relatively low degree of cover, as evidenced by the large pores versus yarn diameters. The yarn bundles are both fine and tightly twisted so that there is little space between their fibers. It is not surprising then to note that the batiste #10 has an air permeability about 5 times that of the polyester textured woven fabric #2, even though their weights per square yard are about equivalent at the 2 oz/yd^2 level. After the cotton fabric #10 is exposed to 200°C, there is little evidence of significant shrinkage or reduction in pore size.

Fabric #11 is an acetate nylon blend, constructed in a warp knit (tricot) fabric. It has a weight of 2.67 oz/yd^2 with 100 yarns/sq.in. The photograph of fabric #11 shows a peculiar roped structure in its surface loops, with considerable separation between the cabled vertical line. The smaller diameter yarns in the back bars of the fabric are just visible in the photograph. It is likely that the yarns visible at the surface are the acetate yarns, while the nylon yarns are visible to the rear of the fabric. The appearance of this fabric would suggest that its porosity, or air permeability, would be rather high, but surprisingly enough, the measured value of 0.4 puts it amongst the less permeable fabrics of the test series. Fabric #11 is said to have been brushed but this is hardly evidenced in the photomicrographs.

Fabric #12 is also a tricot warp-knit fabric, and is constructed of 100% nylon with about the same number of yarns as for fabric #11. However, the weight of fabric #12 is just under

2-1/2 oz/yd^2 somewhat less than fabric #11. The appearance
of fabric #12 in its original state suggests that it would
be a fairly permeable fabric, and this is borne out in measurements which show it to be the most permeable of all the
test fabrics with a test value of 7.4. It is difficult, however,
to sense from the photographs of fabrics #11 and #12 (in the
original state) that fabric #12 is almost 20 times more permeable than fabric #11.

Fabric #12 has been heated to 260° and the resulting
structure was again photographed under the scanning electron
microscope. It would appear that the exposure to 260°C causes
a considerable shrinkage and compaction of fabric #12. It
would be interesting to determine to what extent its extremely
high permeability would be reduced as a result of this temperature exposure. When fabric #12 was exposed to 260°C, some
melting was initiated as is evident in the 300-times enlargement
showing a considerable melt bridging over the fibers in the
yarns of the fabric. In addition, a photomicrograph is provided at 2500-times magnification, indicating a cracking and
splitting of the surface of the nylon fibers as a result of the
thermal exposure. And still another photomicrograph taken of
#12 when exposed to 260°C, shows (at 3000 magnification) melting
and bridging between fibers.

Fabric #13 is an acetate warp knit (continuous filament
yarn) fabric, 2.44 oz/yd^2 in weight with 80 yarns/sq.in. It
is a fairly open fabric with a very high permeability of 6.8.
In fact only fabric #12, the 2.47 oz/yd^2 nylon tricot shows a
higher permeability of 7.4 (ft/sec)/(lb/ft^2).

Another nylon fabric is sample #14, a plain woven taffeta
fabric with 184 yarns/sq.in. and a weight of 1.67 oz/yd^2. The
yarns of fabric #14 are not textured. They are relatively
low-twist continuous filament yarns and during weaving in such
a tight structure, they tend to flatten out, as is evidenced
in the picture of fabric #14 in its original state. One may
contrast the photographs of fabric #14 with the photographs
of the polyester textured woven fabric #2 and note difference

between a flat, continuous filament yarn and a textured, continuous filament yarn as affecting the surface structure of the fabric. It is also constructive to compare fabric #14 with fabric #10. Fabric #14 is a plain weave tightly woven fabric constructed of low-twist continuous filament nylon yarns, having a weight of 1.67 oz/yd^2 whereas fabric #10 is a plain woven fabric made of high-twisted staple cotton yarns with a weight of about 2 oz/yd^2 and constructed with 196 yarns/sq.in. The effect of the higher twist in the cotton yarn fabric #10 is to maintain round yarn bundles resulting in smaller diameters and correspondingly larger pores. In contrast, fabric #14, made of continuous filament low-twist yarns, appears to possess a structure of interlaced ribbons where the low-twist yarns spread out and essentially cover the pores.

When fabric #14 comprised of 100% nylon non-textured yarn is heated to 260°C, the yarns shrink significantly and tend to close up whatever pores existed in the original fabric. Undoubtedly, this results in a further reduction of the relatively low air permeability of 0.58 reported for the original state of fabric #14. This is to be contrasted with the air permeability of 3.4 for fabric #10. The point to be emphasized here is that the technological details provided in Table 2-2 for these two fabrics would suggest that they are essentially of the same construction, in regard to yarns/sq.in. and fabric weight, both with a plain weave. But the nature of the spun yarns vs. the continuous filament yarns is shown to have a very considerable effect on the structure and porosity of the two specimens. In addition, the thermoplastic composition of fabric #14 invites significant shrinkage and fabric tightening effects in high temperature exposure. When fabric #14 is exposed to temperatures of 260°C, the photograph for this condition shows the occurrence of melting at the edge of the specimen despite the tightened integrity of the internal fabric structure. A final photograph shows further the effect of high temperature exposure, at higher magnifications.

Fabric #15 is a woven, durable press slack material in the same weight range as fabrics #1 and #6.
It is durable press treated. However, the yarns in fabric #15 consist of a 65/35 polyester/rayon blend and its weave structure is plain. Despite its heavier yarns and plain (vs. twill) weave the fabric #15 has considerably greater permeability than fabrics #1 and #6 as is suggested by the relative pore sizes shown in the micrographs.

Fabric #16 is a polyester/cotton blend shirting material, with a plain weave. It is considerably lighter than #15 yet with a somewhat higher yarn count, which accounts for its lower permeability. The surface fuzz of cotton fibers is clearly seen in the photomicrograph.

Fabric #17 is a polyester/cotton blend -- a plain weave batiste structure -- similar to the all-cotton batiste, #10. #17 possesses slightly fewer yarns/sq.in., but it is about 1/2 oz/yd^2 heavier than fabric #10, indicating that the individual yarns are somewhat heavier in the blend fabric. The photographs of fabric #17 in its original state show evidence of moderately tightly twisted staple yarn bundles with fairly well defined pores, although some of the pores seem to be covered over either by yarn flattening or by fuzz rising above the yarn structure. The appearance of fabric #17 suggests somewhat less porosity than for the all cotton batiste, fabric #10. This is evidenced in the permeability measurements of 1.89 for the polyester cotton #17 and almost double that for the all cotton batiste #10.

When fabric #17, the polyester cotton, is heated to 260°C, its photographs give the appearance of structural tightening and reduction in pore size. But of more significance is the appearance of considerable bulking in the individual yarns of the fabric. This can take place as a result of shrinkage of the polyester component of the yarn and their tightening down to the center of the bundle. Meanwhile, the cotton fiber, which does not shrink as much, appears to be bulked upward at the surface of the fabric. This type of bulking behavior because of mixed shrinkage characteristics of fibers is well known in textile technology, and has been applied extensively to acrylic staple fabrics made of high shrinkage and low shrinkage fiber. A photograph enlarged to 2500 times magnification is also provided to illustrate the effects of exposure to 230°C on the fibers of fabric #17. The large fibers seen in the lower portion of the photograph appear to be the polyester fibers and these possess very little crimp. There appears to be one cotton fiber in the photograph which to some extent appears to be buckled from the yarn bundle. In addition, there is some evidence of bridging between the fibers due to the initiation of melting.

Fabric #19 is the only fire retardant treated fabric in the GIRCFF series. It is a 3.69 oz/yd^2 all cotton flannel, based on a plain weave. Its yarn frequency is low compared to most of the other fabrics, yet it has one of the lowest permeabilities of the group. The napped nature of the flannel is evident in the fuzz fibers seen in the photomicrograph, yet its surface fiber cover does not approach that of the napped wool flannel, fabric #20.

Fabric #18 is an all cotton flannel, plain weave napped at its surface. It has only 88 yarns/square inch and weight 3.69 ounces/square yard. It is obvious from the photograph of fabric #18 that it is constructed of relatively coarse yarn with large diameters and spread out to cover effectively the pores of the fabric. Thus the air permeability of this structure is on the low side. On the other hand, the yarns appear to be fairly low in twist, as would be needed to develop a good napped surface. Considerable fiber fuzz is evident in the photograph.

Fabric #20 is an all wool flannel with only 57 yarns/square inch and weighing 6.48 oz/yd^2. It is a twill weave napped at its surface. The unique feature of the scanning electron micrograph shown for the original state of fabric

#20, is that there is so much surface fuzz on the fabric that the basic weave cannot be easily observed. This is a clear-cut example where the base fabric would be classified by Bernskiold[12] as being made of staple yarn systems, while the surface would be made of unorganized fiber nap.

Thermophysical Property Summary [R8]
Durable Press Slack
GIRCFF Fabric No. 1

1. Description

   Fiber Composition: 65/35% Polyester/Cotton
   Texture: 141 yarns/sq. in.
   Color: White
   Weave: Twill

2. Mass per Unit Area: 23.49 mg/cm^2; 7.07 oz/yd^2

3. Thickness: 0.0464 cm

4. Specific Heat:

Temperature	50	125	200	°C
Specific Heat	1.19	1.34	1.47	Ws/g C

5. Thermal Conductance

   (i) Contact Pressure   528 N/m^2

	Temperature	63.4	108.6	161.6	°C
	Conductance	15.45	13.31	16.56	m W/cm^{2}K

   (ii) Contact Pressure   866 N/m^2

	Temperature	88.7	°C
	Conductance	15.71	m W/cm^{2}K

6. Ignition Temperature, self 413°C, pilot 334°C

7. Infrared Optical Properties:

	Original		Charred at 219°C
Source:	3,160°K	0.6-2.5µm	3,160°K
Absorptance	0.184	0.189	0.349
Reflectance	0.605	0.522	0.366
Transmittance	0.211	0.289	0.285
Optical Thickness	0.375	0.296	0.490

8. Heat Released in Burning in Air, Yeh [18]:

   (i) Per unit weight   2260 cal/g
   (ii) Per unit area   64 cal/cm^2

9. L.O.I.: 0.171

10. Permeability: 0.18 (ft/sec)/(lb/ft^2)

#1 (ORIGINAL, 85X)

#1 (ORIGINAL, 300X)

Thermophysical Property Summary [R8]

Textured Woven Blouse
GIRCFF Fabric No. 2

1. Description
   Fiber Composition: 100% Polyester
   Texture: 260 yarns/sq. in.
   Color: Yellow
   Weave: Twill

2. Mass Per Unit Area: 7.51 mg/cm^2, 2.19 oz./sq. yd.

3. Thickness: 0.0201 cm

4. Specific Heat

Temperature	50	125	200	°C
Spec. Heat	1.42	1.42	1.42	Ws/g C

5. Thermal Conductance
   (i) Contact Pressure 528 N/m^2

Temperature	73.0	93.1	124.6	°C
Conductance	20.3	21.0	22.8	mW/cm^{2}K

   (ii) Contact Pressure 866 N/m^2

Temperature	83.3	86.3	°C
Conductance	21.0	22.7	mW/cm^{2}K

   (iii) Contact Pressure 1,370 N/m^2

Temperature	75.8	°C
Conductance	22.5	mW/cm^{2}K

6. Melting Temperature   259°C

7. Infrared Optical Properties

	Original		Charred at 134°C
Source:	3,160°K	0.6-2.5μm	3,160°K
Absorptance	0.164	0.153	0.175
Reflectance	0.560	0.501	0.582
Transmittance	0.276	0.346	0.243
Optical Thickness	0.272	0.208	0.305

8. Heat Released in Burning in Air   Yeh [18]
   (i) Per unit weight   n.a.
   (ii) Per unit area   n.a.

9. L.O.I.: 0.190

10. Permeability: 0.70 (ft/sec)(lb/ft^2)

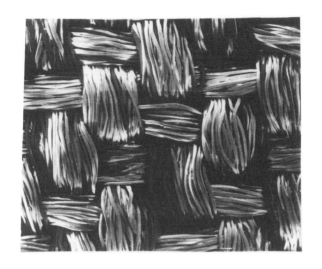

#2 (ORIGINAL, 85X)

#2 (HEATED TO 200°C, 85X)

#2   (HEATED TO 260°C, 100X)

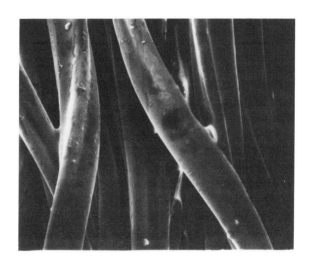

#2   (HEATED TO 230°C, 850X)

Thermophysical Property Summary [R8]

<u>Double Knit</u>
GIRCFF Fabric No. 3

1. <u>Description</u>
   Fiber Composition: 100% Polyester
   Texture: 121 yarns/sq.in.
   Color: White
   Weave: Double-knit

2. <u>Mass Per Unit Area</u>: 20.91 mg/cm^2, 5.83 oz./sq.yd.

3. <u>Thickeness</u>: 0.0953 cm.

4. <u>Specific Heat</u>

Temperature	50	125	200	°C
Spec. Heat	1.03	1.25	1.67	Ws/g C

5. <u>Thermal Conductance</u>
   (i) Contact Pressure   528 N/m^2

Temperature	63.2	109.7	161.0	°C
Conductance	5.40	5.54	6.29	m W/cm^{2}K

   (ii) Contact Pressure   866 N/m^2

Temperature	92.0			°C
Conductance	5.46			m W/cm^{2}K

6. <u>Melting Temperature</u>   251°C

7. <u>Infrared Optical Properties</u>

	Original		Charred at 134°C
Source:	3,160°K	0.6-2.5μm	3,160°K
Absorptance	0.190	0.143	0.200
Reflectance	0.560	0.619	0.506
Transmittance	0.250	0.238	0.294
Optical Thickness	0.335	0.276	0.306

8. <u>Heat Released in Burning in Air</u>:   Yeh [18]
   (i)   Per unit weight   n.a.
   (ii) Per unit area   n.a.

9. <u>L.O.I.</u>: 0.200

10. <u>Permeability</u>: 5.1 (ft/sec) (lb/ft^2)

71

#3 (ORIGINAL, 85X)

#4 (ORIGINAL, 85X)

Thermophysical Property Summary [R8]

Denim
GIRCFF Fabric No. 4

1. Description
    Fiber Composition: 100% Cotton
    Texture: 109 yarns/sq. in.
    Color: Navy Blue
    Weave: Twill

2. Mass Per Unit Area: 29.63 mg/cm^2, 8.66 oz./sq. yd.

3. Thickness: 0.0575 cm

4. Specific Heat

Temperature	50	125	200	°C
Spec. Heat	1.17	1.35	1.50	Ws/g C

5. Thermal Conductance
    (i)   Contact Pressure   528 N/m^2

Temperature	65.2	107.7	159.1	°C
Conductance	9.02	10.88	14.62	mW/cm^{2}K

    (ii) Contact Pressure   866 N/m^2

Temperature	90.2	°C
Conductance	13.05	mW/cm^{2}K

6. Ignition Temperature, self 288°C, pilot 285°C

7. Infrared Optical Properties

	Original		Charred at 162°C
Source:	3,160°K	0.6-2.5μm	3,160°K
Absorptance	0.445	0.372	0.511
Reflectance	0.450	0.491	0.401
Transmittance	0.105	0.137	0.088
Optical Thickness	1.106	0.855	1.305

8. Heat Released in Burning in Air   Yeh [18]
    (i)   Per unit weight: 3200 cal/g
    (ii) Per unit area: 94 cal/cm^2

9. L.O.I.: 0.169

10. Permeability: 0.07 (ft/sec)(lb/ft^2)

73

Thermophysical Property Summary [R8]

<u>T-Shirt, Jersey</u>
GIRCFF Fabric No. 5

1. Description
   Fiber Composition: 100% Cotton
   Texture: 75 yarns/sq. in.
   Color: White
   Weave: Knit

2. <u>Mass Per Unit Area</u>: 13.71 mg/cm^2, 3.97 oz./sq. yd.

3. <u>Thickness</u> 0.0529 cm

4. Specific Heat

Temperature	50	125	200	°C
Spec. Heat	1.05	1.47	1.91	Ws/g C

5. Thermal Conductance
   (i)   Contact Pressure   528 N/m^2

Temperature	70.9	130.5	206.0	°C
Conductance	7.32	8.81	10.82	mW/cm^{2}K

   (ii) Contact Pressure   866 N/m^2

Temperature	57.4	79.8	°C
Conductance	7.31	11.28	mW/cm^{2}K

   (iii) Contact Pressure 1,370 N/m^2

Temperature	76.5	°C
Conductance	13.02	mW/cm^{2}K

6. <u>Ignition Temperature</u>, self 314°C, pilot 306°C

7. <u>Infrared Optical Properties</u>

	Original		Charred at 162°C
Source:	3,160°K	0.6-2.5μm	3,160°K
Absorptance	0.179	0.183	0.225
Reflectance	0.533	0.521	0.491
Transmittance	0.288	0.296	0.284
Optical Thickness	0.284	0.282	0.384

8. <u>Heat Released in Burning in Air</u> Yeh [18]
   (i) Per unit weight: 3160 cal/g
   (ii) Per unit area: 43 cal/cm^2

9. <u>L.O.I.</u>: 0.171

10. <u>Permeability</u>: 1.18 (ft/sec)(lb/ft^2)

74

#5 (ORIGINAL, 85X)

#6 (ORIGINAL, 85X)

Thermophysical Property Summary [R8]

Untreated Slack
GIRCFF Fabric No. 6

1. Description
   Fiber Composition: 65/35% Polyester/Cotton
   Texture: 139 yarns/sq. in.
   Color: White
   Weave: Twill

2. Mass Per Unit Area: 23.57 mg/cm^2, 7.0 oz./yd^2

3. Thickness: 0.0476

4. Specific Heat

Temperature	50	125	200	°C
Spec. Heat	1.10	1.26	1.48	Ws/g C

5. Thermal Conductance
   (i) Contact Pressure 528 N/m^2

Temperature	66.8	12.68	160.6	°C
Conductance	9.90	107.4	15.60	mW/cm^{2}K

   (ii) Contact Pressure 866 N/m^2

Temperature	88.9	°C
Conductance	14.53	mW/cm^{2}K

6. Ignition Temperature, self 412°C, pilot 341°C

7. Infrared Optical Properties

	Original		Charred at 219°C
Source:	3,160°K	0.6-2.5µm	3,160°K
Absorptance	0.199	0.135	0.373
Reflectance	0.581	0.581	0.339
Transmittance	0.220	0.284	0.288
Optical Thickness	0.388	0.223	0.510

8. Heat Released in Burning in Air   Yeh [18]
   (i) Per unit weight: 2790 cal/g
   (ii) Per unit area: 67 cal/cm^2

9. L.O.I.: 0.171

10. Permeability: 0.18 (ft/sec)(lb/ft^2)

Thermophysical Property Summary [R8]

Jersey Tube Knit
GIRCFF Fabric No. 7

1. Description
   Fiber Composition: 100% Acrylic
   Texture: 52 yarns/sq. in.
   Color: Gold
   Weave: Knit

2. Mass Per Unit Area: 15.13 mg/cm^2, 2.64 oz./sq. yd.

3. Thickness: 0.0791 cm

4. Specific Heat

Temperature	50	125	200	°C
Spec. Heat	1.27	1.71		Ws/g C

5. Thermal Conductance
   (i) Contact Pressure 528 N/m^2

Temperature	66.4	106.6	142.7	°C
Conductance	5.44	8.66	9.92	mW/cm^{2}K

   (ii) Contact Pressure 866 N/m^2

Temperature	88.3	°C
Conductance	7.93	mW/cm^{2}K

6. Melting Temperature   241°C

7. Infrared Optical Properties

	Original		Charred at 230°C
Source:	3,160°K	0.6-2.5µm	3,160°K
Absorptance	0.202	0.292	0.720
Reflectance	0.536	0.549	0.113
Transmittance	0.262	0.159	0.167
Optical Thickness	0.340	0.855	1.115

8. Heat Released in Burning in Air Yeh [18]
   (i) Per unit weight   n.a.
   (ii) Per unit area: n.a.

9. L.O.I.: 0.162

10. Permeability: 2.6 (ft/sec)(lb/ft^2)

77

#7 (ORIGINAL, 85X)

#8 (ORIGINAL, 85X)

Thermophysical Property Summary [R8]

T-Shirt, Jersey
GIRCFF Fabric No. 8

1. Description
    Fiber Composition: 65/35% Polyester/Cotton
    Texture: 77 yarns/sq. in.
    Color: White
    Weave: Knit

2. Mass Per Unit Area: 16.19 mg/cm^2, 4.8 oz/yd^2

3. Thickness: 0.062 cm

4. Specific Heat

Temperature	50	125	200	°C
Spec. Heat	1.23	1.54	1.86	Ws/g C

5. Thermal Conductance
    (i)   Contact Pressure 528 N/m^2

Temperature	68.4	129.6	205.9	°C
Conductance	7.63	10.50	11.13	mW/cm^{2}K

    (ii) Contact Pressure 866 N/m^2

Temperature	56.0	81.7	°C
Conductance	7.99	9.13	mW/cm^{2}K

   (iii) Contact Pressure 1,370 N/m^2

Temperature	75.4	°C
Conductance	10.74	mW/cm^{2}K

6. Ignition Temperature, self 437°C, pilot 310°C

7. Infrared Optical Properties

	Original		Charred at 230°C
Source:	3,160°K	0.6-2.5μm	3,160°K
Absorptance	0.174	0.164	0.346
Reflectance	0.562	0.560	0.346
Transmittance	0.264	0.276	0.308
Optical Thickness	0.298	0.272	0.460

8. Heat Released in Burning in Air   Yeh [18]
    (i)   Per unit weight: 3160 cal/g
    (ii) Per unit area: 51 cal/cm^2

9. L.O.I.: 0.167

10. Permeability: 1.03 (ft/sec)(lb/ft^2)

79

Thermophysical Property Summary [R8]

Terry Cloth
GIRCFF Fabric No. 9

1. Description
   Fiber Composition: 100% Cotton
   Texture: 87 yarns/sq. in.
   Color: White
   Weave: Pile

2. Mass Per Unit Area: 26.48 mg/cm^2, 7.42 oz/sq. yd.

3. Thickness: 0.2080 cm

4. Specific Heat

Temperature	50	125	200	°C
Spec. Heat	1/13	1.45	1.64	Ws/g C

5. Thermal Conductance
   (i) Contact Pressure 528 N/m^2

Temperature	69.7	117.4	175.3	°C
Conductance	2.34	2.79	3.62	mW/cm^{2}K

   (ii) Contact Pressure 866 N/m^2

Temperature	95.3	°C
Conductance	2.98	mW/cm^{2}K

6. Ignition Temperature, self 302°C, pilot 294°C

7. Infrared Optical Properties

	Original		Charred at 162°C
Source:	3,160°K	0.6-2.5µm	3,160°K
Absorptance	0.113	0.146	0.242
Reflectance	0.681	0.623	0.627
Transmittance	0.206	0.231	0.131
Optical Thickness	0.274	0.288	0.660

8. Heat Released in Burning in Air   Yeh [18]
   (i) Per unit weight: n.a.
   (ii) Per unit area: n.a.

9. L.O.I.: 0.158

10. Permeability: 1.20 (ft/sec)(lb/ft^2)

#9 (ORIGINAL, FRONT SIDE, 25X)

#9 (ORIGINAL, BACK SIDE, 25X)

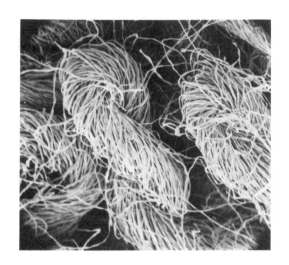

#9 (ORIGINAL, FRONT SIDE, 45X)

#9 (ORIGINAL, BACK SIDE, 45X)

Thermophysical Property Summary [R8]

Batiste
GIRCFF Fabric No. 10

1. Description
    Fiber Composition: 100% Cotton
    Texture: 187-196 yarns/sq. in.
    Color: Purple
    Weave: Plain

2. Mass Per Unit Area: 6.65 mg/cm^2, 1.90-2.04 oz./sq. yd.

3. Thickness: 0.0177 cm

4. Specific Heat

Temperature	50	125	200	°C
Specific Heat	1.37	1.77	--	Ws/g C

5. Thermal Conductance
    (i) Contact Pressure 528 N/m^2

Temperature	67.8	128.4	195.8	°C
Conductance	17.48	20.19	29.67	mW/cm^{2}K

    (ii) Contact Pressure 1,370 N/m^2

Temperature	76.5	°C
Conductance	25.36	mW/cm^{2}K

6. Ignition Temperature, self 437°C, pilot 345°C

7. Infrared Optical Properties

	Original		Charred at 230°C
Source:	3,160°K	0.6-2.5μm	3,160°K
Absorptance	0.221	0.200	0.387
Reflectance	0.418	0.406	0.219
Transmittance	0.361	0.394	0.394
Optical Thickness	0.280	0.236	0.413

8. Heat Released in Burning in Air   Yeh [18]
    (i) Per unit weight: 3120 cal/g
    (ii) Per unit area: 20 cal/cm^2

9. L.O.I.: 0.162

10. Permeability: 3.4 (ft/sec)(lb/ft^2)

#10   (ORIGINAL, 85X)

#10   (HEATED TO 200°C, 85X)

Thermophysical Property Summary [R8]

Tricot
GIRCFF Fabric No. 11

1. Description
   Fiber Composition: 80/20% Acetate/Nylon
   Texture: 100 yarns/sq. in.
   Color: White
   Weave: Knit

2. Mass Per Unit Area: 11.31 mg/cm^2

3. Thickness: 0.0614 cm, 2.67 oz./sq. yd.

4. Specific Heat

Temperature	50	125	200	°C
Spec. Heat	1.45	1.45	1.45	Ws/g C

5. Thermal Conductance
   (i)   Contact Pressure 528 N/m^2

Temperature	75.2	95.2	130.9	°C
Conductance	6.60	7.08	8.08	mW/cm^{2}K

   (ii) Contact Pressure 866 N/m^2

Temperature	86.2	°C
Conductance	7.46	mW/cm^{2}K

   (iii) Contact Pressure 1,370 N/m^2

Temperature	77.7	°C
Conductance	8.13	mW/cm^{2}K

6. Melting Temperature   228°C

7. Infrared Optical Properties

	Original		Charred at 134°C
Source:	3,160°K	0.6-2.5µm	3,160°K
Absorptance	0.157	0.163	0.178
Reflectance	0.523	0.508	0.566
Transmittance	0.320	0.329	0.256
Optical Thickness	0.230	0.232	0.303

8. Heat Released in Burning in Air Yeh [18]
   (i)   Per unit weight: n.a.
   (ii) Per unit area: n.a.

9. L.O.I.: 0.179

10. Permeability: 0.40 (ft/sec)(lb/ft^2)

#11 (ORIGINAL, 85X)

Thermophysical Property Summary [R8]

Tricot
GIRCFF Fabric No. 12

1. Description
   Fiber Composition: 100% Nylon
   Texture: 98 yarns/sq. in.
   Color: White
   Weave: Knit

2. Mass Per Unit Area: 8.91 mg/cm^2, 2.47 oz./sq. yd.

3. Thickness 0.0271 cm

4. Specific Heat

Temperature	50	125	200	°C
Spec. Heat	1.51	2.09	2.68	Ws/g C

5. Thermal Conductance
   (i)   Contact Pressure   528 N/m^2

Temperature	76.1	96.3	122.6	°C
Conductance	19.3	17.00	20.93	mW/cm^{2}K

   (ii) Contact Pressure   866 N/m^2

Temperature	56.1	92.2	°C
Conductance	17.35	20.8	mW/cm^{2}K

   (iii) Contact Pressure 1,370 N/m^2

Temperature	73.5	°C
Conductance	17.89	mW/cm^{2}K

6. Melting Temperature   246°C

7. Infrared Optical Properties

	Original		Charred at 134°C
Source:	3,160°K	0.6-2.5µm	3,160°K
Absorptance	0.173	0.170	0.189
Reflectance	0.447	0.397	0.465
Transmittance	0.380	0.433	0.346
Optical Thickness	0.214	0.188	0.289

8. Heat Released in Burning in Air   Yeh [18]
   (i)   Per Unit Weight: n.a.
   (ii) Per Unit Area: n.a.

9. L.O.I.: 0.198

10. Permeability: 7.4 (ft/sec)(lb/ft^2)

#12 (ORIGINAL, 85X)

#12 (HEATED TO 260°C, 85X)

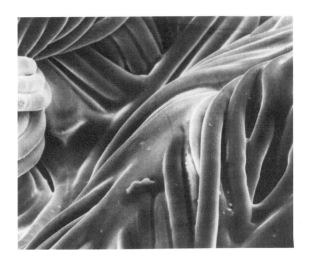

#12   (HEATED TO 260°C, 300X)

#12   (HEATED TO 260°C, 2500X)

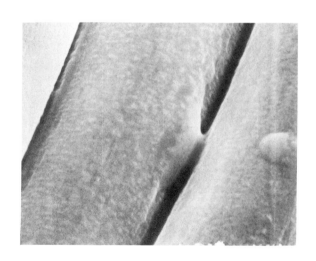

#12 (HEATED TO 260°C, 3000X)

Thermophysical Property Summary [R8]

Tricot
GIRCFF Fabric No. 13

1. Description
   Fiber Composition: 100% Acetate
   Texture: 80 yarns/sq. in.
   Color: White
   Weave: Knit

2. Mass Per Unit Area: 9.40 mg/cm^2, 2.44 oz./sq. yd.

3. Thickness: 0.0304 cm

4. Specific Heat

Temperature	50	125	200	°C
Spec. Heat	1.24	1.65	2.05	Ws/g C

5. Thermal Conductance
   (i)   Contact Pressure 528 N/m^2

Temperature	72.1	94.9	125.1	°C
Conductance	12.21	13.93	15.86	mW/cm^{2}K

   (ii) Contact Pressure 866 N/m^2

Temperature	91.7	°C
Conductance	18.82	mW/cm^{2}K

   (iii) Contact Pressure 1,370 N/m^2

Temperature	76.9	°C
Conductance	17.56	mW/cm^{2}K

6. Melting Temperature   227°C

7. Infrared Optical Properties

	Original		Charred at 134°C
Source:	3,160°K	0.6-2.5μm	3,160°K
Absorptance	0.166	0.167	0.182
Reflectance	0.490	0.443	0.492
Transmittance	0.344	0.390	0.326
Optical Thickness	0.227	0.203	0.291

8. Heat Released in Burning in Air   Yeh [18]
   (i)   Per unit weight: n.a.
   (ii) Per unit area: n.a.

9. L.O.I.: 0.179

10. Permeability: 6.8 (ft/sec)(lb/ft^2)

91

#13 (ORIGINAL, 85X)

Thermophysical Property Summary [R8]

Taffeta
GIRCFF Fabric No. 14

1. Description
   Fiber Composition: 100% Nylon
   Texture: 184 yarns/sq. in.
   Color: White
   Weave: Plain

2. Mass Per Unit Area: 5.66 mg/cm^2, 1.67 oz./sq. yd.

3. Thickness: 0.0110 cm

4. Specific Heat

Temperature	50	125	200	°C
Spec. Heat	1.59	1.69	2.13	Ws/g C

5. Thermal Conductance
   (i) Contact Pressure 528 N/m^2

Temperature	63.2	104.4	156.4	°C
Conductance	33.91	35.71	41.91	mW/cm^{2}K

   (ii) Contact Pressure 866 N/m^2

Temperature	88.2	°C
Conductance	27.80	mW/cm^{2}K

6. Melting Temperature   241°C

7. Infrared Optical Properties

	Original		Charred at 134°C
Source:	3,160°K	0.6-2.5µm	3,160°K
Absorptance	0.144	0.218	0.193
Reflectance	0.348	0.348	0.399
Transmittance	0.508	0.434	0.408
Optical Thickness	0.136	0.234	0.222

8. Heat Released in Burning in Air   Yeh [18]
   (i) Per unit weight: n.a.
   (ii) Per unit area: n.a.

9. L.O.I.: 0.247

10. Permeability: 0.58 (ft/sec)(lb/ft^2)

#14   (ORIGINAL, 85X)

#14   (HEATED TO 260°C, 100X)

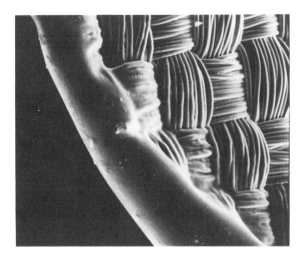

#14   (HEATED TO 260°C, 100X)

#14   (HEATED TO 260°C, 300X)

Thermophysical Property Summary [R8]

Durable Press Slack

GIRCFF Fabric No. 15

1. Description

Fiber Composition: 65/35% Polyester
Texture: 91 yarns/sq. in.
Color: Brown
Weave: Plain

2. Mass Per Unit Area: 22.82 mg/cm^2, 6.62 oz/yd^2

3. Thickness: 0.0421 cm

4. Specific Heat

Temperature	50	125	200	°C
Specific Heat	0.91	1.26	1.55	Ws/g C

5. Thermal Conductance

(i) Contact Pressure 528 N/m^2

Temperature	64.1	104.7	159.8	°C
Conductance	12.01	13.91	16.91	mW/cm^{2}K

(ii) Contact Pressure 866 N/m^2

Temperature 89.0 °C
Conductance 15.37 mW/cm^{2}K

6. Ignition Temperature, self 432°C, pilot 334°C

7. Infrared Optical Properties

	Original		Charred at 247°C
Source:	3,160°K	0.6-2.5μm	3,160°K
Absorptance	0.226	0.256	0.603
Reflectance	0.515	0.413	0.239
Transmittance	0.259	0.331	0.158
Optical Thickness	0.375	0.340	1.040

8. Heat Released in Burning in Air. Yeh [18]

(i) Per unit weight: 2400 cal/g
(ii) Per unit area: 54 cal/cm^2

9. L.O.I.: 0.179

10. Permeability: 1.71 (ft/sec) (lb/ft^2)

96

#15 (ORIGINAL, 85X)

#16 (ORIGINAL, 85X)

Thermophysical Property Summary [R8]

Shirting
GIRCFF Fabric No. 16

1. Description
   Fiber Composition: 50/50% Polyester/Cotton
   Texture: 102 yarns/sq. in.
   Color: White
   Weave: Plain

2. Mass Per Unit Area: 13.14 mg/cm^2, 3.91 oz./sq. yd.

3. Thickness: 0.0321 cm

4. Specific Heat

Temperature	50	125	200	°C
Spec. Heat	1.16	1.31	1.44	Ws/g C

5. Thermal Conductance
   (i)   Contact Pressure   528 N/m^2

Temperature	63.9	105.9	135.6	°C
Conductance	14.49	14.62	19.03	mW/cm^{2}K

   (ii) Contact Pressure   866 N/m^2
         Temperature 90.7   °C
         Conductance 19.72   mW/cm^{2}K

6. Ignition Temperature, self 470°C, pilot 357°C

7. Infrared Optical Properties

	Original		Charred at 247°C
Source:	3,160°K	0.6-2.5μm	3,160°K
Absorptance	0.168	0.207	0.241
Reflectance	0.613	0.499	0.322
Transmittance	0.219	0.294	0.437
Optical Thickness	0.338	0.312	0.256

8. Heat Released in Burning in Air   Yeh [18]
   (i)   Per unit weight: 2780 cal/g
   (ii) Per unit area: 37 cal/cm^2

9. L.O.I.: 0.182

10. Permeability: 0.92 (ft/sec)(lb/ft^2)

Thermophysical Property Summary [R8]

Batiste
GIRCFF Fabric No. 17

1. Description
    Fiber Composition: 65/35% Polyester/Cotton
    Texture: 173 yarns/sq. in.
    Color: White
    Weave: Plain

2. Mass Per Unit Area: 8.55 mg/cm^2, 2.53 oz./sq. yd.

3. Thickness: 0.0193 cm

4. Specific Heat

Temperature	50	125	200	°C
Spec. Heat	1.27	1.47	1.66	Ws/g C

5. Thermal Conductance
    (i) Contact Pressure 528 N/m^2

Temperature	72.6	128.0	199.3	°C
Conductance	17.34	22.48	26.75	mW/cm^{2}K

    (ii) Contact Pressure 1,370 N/m^2

Temperature	77.3	°C
Conductance	30.41	mW/cm^{2}K

6. Ignition Temperature, self 471°C, pilot 376°C

7. Infrared Optical Properties

	Original		Charred at 252°C
Source:	3,160°K	0.6-2.5μm	3,160°K
Absorptance	0.151	0.164	0.425
Reflectance	0.485	0.464	0.255
Transmittance	0.364	0.372	0.320
Optical Thickness	0.198	0.208	0.520

8. Heat Released in Burning in Air   Yeh [18]
    (i) Per unit weight: 2970 cal/g
    (ii) Per unit area: 25 cal/cm^2

9. L.O.I.: 0.179

10. Permeability: 1.89 (ft/sec)(lb/ft^2)

99

#17 (ORIGINAL, 85X)

#17 (HEATED TO 260°C, 100X)

#17   (HEATED TO 230°C, 2500X)

Thermophysical Property Summary [R8]

Flannel
GIRCFF Fabric No. 18

1. Description
    Fiber Composition: 100% Cotton
    Texture 88 yarns/sq. in.
    Color: White
    Weave: Plain

2. Mass Per Unit Area: 12.88 mg/cm^2, 3.69 oz./sq. yd.

3. Thickness: 0.0713 cm

4. Specific Heat

Temperature	50	125	200	°C
Spec. Heat	1.44	1.63	1.82	Ws/g C

5. Thermal Conductance
    (i)   Contact Pressure   528 N/m^2

Temperature	73.0	134.4	205.3	°C
Conductance	5.55	6.72	8.74	mW/cm^{2}K

    (ii) Contact Pressure   1,370 N/m^2
          Temperature 78.6   °C
          Conductance 7.06   mW/cm^{2}K

6. Ignition Temperature, self 306°C, pilot 286°C

7. Infrared Optical Properties

	Original		Charred at 162°C
Source:	3,160°K	0.6-2.5µm	3,160°K
Absorptance	0.170	0.176	0.241
Reflectance	0.599	0.573	0.527
Transmittance	0.231	0.251	0.232
Optical Thickness	0.326	0.312	0.431

8. Heat Released in Burning in Air   Yeh [18]
    (i)   Per unit weight: 3270 cal/g
    (ii) Per unit area: 41 cal/cm^2

9. L.O.I.: 0.171

10. Permeability: 1.19 (ft/sec)(lb/ft^2)

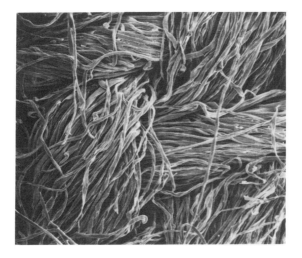

#18　(ORIGINAL, 85X)

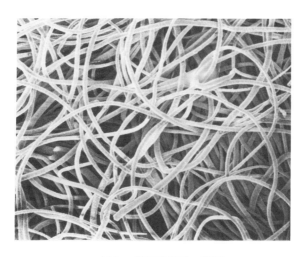

#20　(ORIGINAL, 85X)

Thermophysical Property Summary [R8]

Flannel
GIRCFF Fabric No. 19

1. Description
   Fiber Composition: 100% Cotton, Fire retardant
   Texture: 89 yarns/sq. in.
   Color: White
   Weave: Plain

2. Mass Per Unit Area: 14.89 mg/cm^2, 4.21 oz/sq. yd.

3. Thickness: 0.0692 cm

4. Specific Heat

Temperature	50	125	200	°C
Spec. Heat	1.33	1.60	1.87	Ws/g °C

5. Thermal Conductance
   (i) Contact Pressure   528 N/m^2

Temperature	73.0	134.1	203.5	°C
Conductance	5.38	7.55	9.46	mW/cm^{2}K

   (ii) Contact Pressure 1,370 N/m^2

Temperature	77.8	°C
Conductance	8.55	mW/cm^{2}K

6. Ignition Temperature, self 488°C, pilot 323°C

7. Infrared Optical Properties

	Original		Charred at 252°C
Source:	3,160°K	0.6-2.5μm	3,160°K
Absorptance	0.181	0.201	0.498
Reflectance	0.590	0.602	0.248
Transmittance	0.229	0.197	0.254
Optical Thickness	0.348	0.428	0.688

8. Heat Released in Burning in Air   Yeh [18]
   (i) Per unit Weight: n.a.
   (ii) Per unit Area: n.a.

9. L.O.I.: 0.252

10. Permeability: 0.09 (ft/sec)(lb/ft^2)

#19 (ORIGINAL, 85X)

Thermophysical Property Summary [R8]

Flannel
GIRCFF Fabric No. 20

1. Description
   Fiber Composition: 100% Wool
   Texture: 54 yarns/sq. in.
   Color: Navy Blue
   Weave: Twill

2. Mass Per Unit Area: 19.93 mg/cm^2, 6.48 oz./sq/ yd.

3. Thickness: 0.0721 cm

4. Specific Heat

Temperature	50	125	200	°C
Spec. Heat	1.27	1.27	1.27	Ws/g C

5. Thermal Conductance
   (i) Contact Pressure   528 N/m^2

	Temperature	63.9	107.5	142.5	°C
	Conductance	7.12	7.29	7.55	mW/cm^{2}K

   (ii) Contact Pressure   866 N/m

	Temperature	91.7	°C
	Conductance	7.00	mW/cm^{2}K

6. Ignition Temperature, self 471°C, pilot 322°C

7. Infrared Optical Properties

	Original		Charred at 252°C
Source:	3,160°K	0.6-2.5µm	3,160°K
Absorptance	0.545	0.533	0.609
Reflectance	0.336	0.365	0.105
Transmittance	0.119	0.102	0.286
Optical Thickness	1.149	1.226	0.725

8. Heat Released in Burning in Air   Yeh [18]

   (i) Per unit Weight: n.a.
   (ii) Per unit Area: n.a.

9. L.O.I.: 0.261

10. Permeability: 0.92 (ft/sec)(lb/ft^2)

CHAPTER III

IGNITION

3.1   INTRODUCTION   108

       Fabric Ignition   109
       Ignition Versus Melting   110

3.2   HEAT TRANSFER ANALYSES   111

       3.2.1 Thermally Thin Slab   112
       3.2.2 Thermally Thick Slab   116

3.3   PRE-IGNITION HEAT TRANSFER THROUGH A FABRIC
       SKIN SIMULANT SYSTEM   124

       3.3.1 Experimental Procedure   125
       3.3.2 Experimental and Theoretical Results   125

3.4   RADIATIVE IGNITION   131

       3.4.1 Experimental Procedure at GIT   131
       3.4.2 Results   131
       3.4.3 Discussion of Results   136
       3.4.4 Proposed Ranking of GIRCFF Fabrics As
              To Ease Of Ignition Or Melting
              (Overview Team)   137
       3.4.5 Summary of Results of Ranking   151

3.5   FLAMING IGNITION   152

       3.5.1 Experimental Procedures   152
       3.5.2 Results and Discussion   161
       3.5.3 Proposed Ranking of GIRCFF Fabrics As
              To Ease of Ignition Or Melting
              (Overview Team)   168

3.6   COMPARISON OF RADIATIVE AND FLAMING IGNITION   174

3.7   SUMMARY OF RESULTS   182

CHAPTER III

IGNITION

3.1 INTRODUCTION

The main objective of the ignition studies under the GIRCFF
Fabric Flammability Program was to determine the probability of
ignition of a fabric (or garment), given a specific thermal
exposure.

Factory Mutual Research Corporation (FMRC) has shown that
ignition for the GIRCFF fabrics occurs well before the wearer
of garments comprised of such fabrics would feel any pain due
to heat transfer from a heat source through the fabrics[R9-R13].
Consequently, pain-activated evasive action (by the wearer)
seems improbable before ignition.

To assess the relative probability of ignition associated
with the selected GIRCFF fabrics, Georgia Institute of
Technology (GIT) measured their ignition times or melting times
under specified heat fluxes from radiative and flaming heat
sources[TR1-R8]. Also, attempts were made to predict theoretically these ignition or melting times. To compare experimental
and theoretical results, GIT determined the physical and thermal
properties of the 20 GIRCFF fabrics. This provided the most
extensive collection of thermophysical data on textiles reported
to date.

In their studies of the GIRCFF fabrics, FMRC[R14-R17]
determined minimum ignition times or melting times under
conditions of flaming ignition. Furthermore, FMRC investigated
the dependence of these minimum times on fabric variables (such
as weight and porosity), flame variables (such as burner size,
fuel rate, air/fuel ratio), presence of a skin simulant and
sample orientation in the gravity field.

To model more realistic ignition hazards, the Gillette
Research Institute (GRI) determined ignition times for sleeve-
like GIRCFF fabric specimens exposed to a gas stove and

investigated the dependence of the ignition times on fiber content of the fabric, on fabric structure, relative humidity and location above the stove [R27].

Using the results obtained in the GIRCFF studies the M.I.T. Overview Team proposed a set of parameters to rank the GIRCFF fabrics with respect to heat input necessary for ignition or melting. These parameters depend on the thermal and physical properties of the fabrics and have potential value as a guideline in the design of fabrics with longer ignition times or longer melting times and hence a lower probability of ignition (or melting)-- given a specific thermal exposure.

Fabric Ignition

When subjected to thermal exposure, cellulosic fabrics heat up and undergo in-depth thermal decomposition.[*] Combustible vapors are generated and chars are formed. When fabric temperatures reach the so called "ignition temperature", run-away exothermic reactions are triggered - accompanied by the appearance of a flame or a glowing zone. This phenomenon is called "ignition" and the time interval between the onset of heating and ignition is called the "ignition time."

Ignition is a complex phenomenon involving processes such as heat-transfer and thermal decomposition governed by fluid mechanics and chemical kinetics [25]. For synthetic fabrics, the thermoplastic behavior (described below) adds to the complexity of the phenomenon. To develop a full understanding of fabric ignition is a formidable task and must be the goal for future research. However, even without an understanding of ignition, knowledge of ignition times provides some indication of the likelihood of fabric ignition under specified conditions and can serve a useful purpose in reducing the hazards of clothing fires. Thus, in this study, emphasis is placed on measuring

---

[*]According to Kilzer and Broido [24], cellulose starts to decompose in the temperature range 200 to 400°C.

ignition times and on determining their dependence on fabric
and non-fabric variables. These experimental results are
then compared to theoretically determined times of heating
to ignition temperatures on the basis of heat-transfer analyses
of inert fabrics (i.e. with negligible thermal decomposition).
Radiative and flaming ignition[*] are treated separately because
the governing heat-transfer mechanisms are different in each
case.

Ignition Versus Melting

The ignition behavior of cellulosic fabrics has been described above. The behavior of thermoplastic fabrics upon
heating is quite different as visual observations of these
fabrics under thermal exposure indicate that melting and surface-
vaporization occur well before ignition (i.e., the appearance of
a flame). Shrinking and dripping may also occur before (and
after) the appearance of the flame. For cellulosic/thermoplastic
blend fabrics, these processes occur to different extents
depending on the relative composition of the fiber blend. Thus,
depending on the fiber content of a fabric, its response to
heating is either ignition or melting (followed by ignition
if heating continues).

Ignition and melting are intrinsically different phenomena
and their occurrence implies different hazards. For example,
ignition is usually followed by flame spreading and the fire may
go out of control; however, melting will cause only localized
burns unless external heating continues, leading to ignition.
Then, flame spreading may occur. Unfortunately, in the GIRCFF
program, ignition and melting were identified as separate physical
entities, yet, they were equated in indicating the onset of "hazard".
This is illustrated by the definitions used by FMRC and GIT in
reporting their results. FMRC defined a "response time" to

---

[*] Radiative ignition sources include electric heaters and hot
plates. Flaming ignition sources include gas ranges, gas space
heaters, matches and lighters. They are involved in about 10%
and 40% of fabric-fire accidents, respectively [26].

denote the onset of ignition or the destruction of one yarn
(i.e. by melting), depending on whether the fabric ignites
or melts as a first response to a thermal flux; and GIT
defined a "destruction time" to denote the occurrence of
ignition or melt-through of the fabric.

It is interesting to note, however, that FMRC reported
some "ignition times" for thermoplastic fabrics (e.g. Table
3-5 in Section 3.5.2) - denoting the appearance of a flame
after the fabric started to melt. It is believed that such
ignition times are to be preferred to melting times in comparing the ignition results for fabrics of different fiber
content.

3.2 HEAT TRANSFER ANALYSES

In the previous section consideration was given to the
complexities of the ignition phenomenon. These complexities
were studied at length by the GIT group. GIT established theoretical models to describe the processes of fabric ignition,
including heating (by radiation or convection), cooling (by
radiation and convection), conduction, energy storage, rate
processes (namely moisture desorption and gasification) and
energy feedback from the gas phase to the solid phase. The GIT
model yielded sets of non-linear differential equations whose
solutions were a formidable task. These equations were solved
for at least one fabric example.

The GIT analyses showed that conduction within the fabric
plays an insignificant role [R4]. This permits appreciable
simplifications in the governing equations. In addition, the
GIT models were progressively simplified and the resulting
equations made even more tractable [R4 and R8].

In lieu of detailing the overall GIT models, we present
stripped down models found useful in comparing theory with
experiments on actual GIRCFF fabrics. The following analysis
prepared by the Overview Team is drawn from classical
textbook treatments of heat transfer. It is based
on simple physical premises and its results are in

reasonable agreement with the simplified versions of the GIT treatments.

The following material is intended to introduce the textile technologist and fiber scientist to the standard heat transfer analyses for cases relevant to fabric ignition. In all instances the fabric is considered to be inert, that is, with negligible endothermic or exothermic reactions before the occurrence of ignition or melting; and the flow of thermal energy to and from the cloth is then equated to its change in internal thermal energy. The fine structure of the fabric is neglected, i.e. its yarn structure, weave, porosity, etc., and it is considered an opaque solid. Further, it is classified as a thermally thin slab with no sub-surface temperature gradients, or as a thermally thick slab with significant thermal gradients. The analyses show that thermal behavior relating to ignition is different for the two cases named.

3.2.1 Thermally Thin Slab

As illustrated in Fig. 3-1, an energy balance over a unit area of a thermally thin slab, heated with a constant radiative flux, W_o, cooled by convection, yields:

$$\text{Energy Increase} = \text{Heat Absorbed}^* - \text{Heat Loss}$$

or $\qquad \rho c \delta \dfrac{dT}{d\tau} = \tilde{\alpha}\, W_o - h(T - T_o)$ $\qquad\qquad$ (3-1)

where $\quad$ T = temperature, °C

$\qquad\qquad \tau$ = time, seconds

$\qquad\qquad \rho$ = density, g/cm^3

$\qquad\qquad$ c = specific heat, $W \cdot s/g \cdot °C$

$\qquad\qquad \delta$ = thickness, cm

$\qquad\qquad$ h = convective heat-transfer coefficient, $W/cm^2 °C$

$\qquad\qquad \tilde{\alpha}$ = absorptivity

and the subscript o denotes initial conditions.

---

*It is assumed that the slab absorbs only a fraction ($\tilde{\alpha}$) of the impinging heat flux (W_o).

112

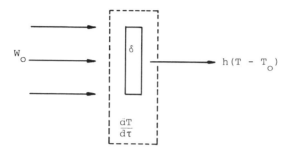

FIGURE 3-1   ENERGY BALANCE OVER A THERMALLY THIN SLAB

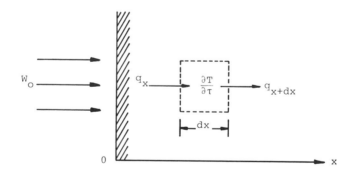

FIGURE 3-2   SEMI-INFINITE SLAB

113

The cases of negligible and non-negligible heat losses are considered separately below:

A. Case with Negligible Heat Losses

The case with negligible heat loss through convection is realized if the third term in eq. (3-1) above is negligible. This occurs if the convective heat transfer coefficient, h, is zero. Setting h = 0 in eq. (3-1) and integrating over the ignition (or melting) time interval, $\tau_{i,m}$, yields:

$$\rho \delta c (T_{i,m} - T_o) = \tilde{\alpha} W_o \cdot \tau_{i,m} \qquad (3-2)$$

where the subscripts i and m denote ignition and melting, respectively.
The Overview Team now sets:

$$(N_{Fo})_{i,m} \equiv \frac{(k/\delta) \cdot \tau_{i,m}}{\rho \delta c} \qquad (3-3)$$

and

$$q^*_{rad} \equiv \frac{\tilde{\alpha} W_o}{(k/\delta)(T_{i,m} - T_o)} \qquad (3-4)$$

In eq. (3-3) the physical time to ignition or melting, i.e. $\tau_{i,m}$, is normalized by dividing by the thermal diffusion time, $\rho \delta c/(k/\delta)$ which is the ratio of thermal capacitance for a unit area to thermal conductance. The unit of this ratio can be easily checked as that of time and it should be noted that thermal diffusion time is the time needed for a thermal wave to diffuse into a material for a distance δ.

Similarly, the absorbed heat flux ($\tilde{\alpha} W_o$) is normalized by dividing by the flux $(k/\delta)(T_{i,m} - T_o)$

114

which depends on the thermophysical properties of the
fabric. Thus, eq. (3-2) can be written in dimension-
less form as:

$$(N_{Fo})_{i,m} = (q^*_{rad})^{-1} \tag{3-2a}$$

It should be noted that the thermal conductance,
k/δ, is introduced in eqs. (3-3) and (3-4) simply to
enable a comparison of the results for thermally thin
and thermally thick slabs. This will be evident below.

In flaming ignition, the absorbed radiative heat
flux, $\tilde{\alpha}W_o$, is replaced by the absorbed convective heat
flux, $h_f \cdot \Delta T_f$,

where h_f = convective heat transfer coefficient
        from flame to fabric, $W/cm^2 \,^{\circ}C$

and

$\Delta T_f = T_f - T_{avg}$ = excess flame temperature over
            average temperature, both
            assumed to be constant over the
            ignition (or melting) time
            interval.

Then, eq. (3-2) becomes:

$$\rho \delta c (T_{i,m} - T_o) = h_f \cdot \Delta T_f \cdot \tau_{i,m} \tag{3-5}$$

If we let:

$$q^{**}_c \equiv \frac{h_f \cdot \Delta T_f}{(k/\delta)(T_{i,m} - T_o)} \tag{3-6}$$

Then eq. (3-5) becomes:

$$(N_{Fo})_{i,m} = (q^{**}_c)^{-1} \tag{3-5a}$$

Note that this is only an approximate relation because
(ΔT_f) actually varies over the ignition interval.

B. Case With Significant Heat Losses

For the case where convective heat losses are significant, it follows that the heat transfer co-efficient (h) will have a significant value.

Eq. (3-1) with h ≠ 0 can be integrated, yielding:

$$\frac{(T_{i,m} - T_o)h}{\tilde{\alpha} W_o} = 1-\exp\left(-\frac{h \cdot \tau_{i,m}}{\rho \delta c}\right) \qquad (3-7)$$

Defining the ratio of the convective coefficient over the conductive as the Biot Number, $N_{Bi} \equiv \frac{h}{k/\delta}$, Eq. (3-7) becomes in non-dimensional form:

$$(N_{Fo})_{i,m} = -\frac{1}{N_{Bi}} \ln\left(1 - \frac{N_{Bi}}{q^*_{rad}}\right) \qquad (3-7a)$$

This equation in effect expresses the normalized ignition (or melting) time as a function of the normalized heating intensity and shows the effect of the relative strength of the convective coefficient on the ignition (or melting) time.

3.2.2 Thermally Thick Slab

For a small element of unit area and of thickness dx within a thermally thick slab, the rate of increase of internal thermal energy equals the net inflow of heat into the element (see Fig. 3-2):

$$\frac{\partial}{\partial \tau} (\rho c T dx) = q_x - q_{x+dx} \qquad (3-8)$$

where q_x = conductive heat flux in the x-direction at position x, W/cm^2.

116

For constant ρ and c, eq. (3-8) becomes:

$$\rho c \ \frac{\partial T}{\partial \tau} = - \ \frac{\partial q}{\partial x} \tag{3-8a}$$

For a unit area, the conductive heat flux is proportional to the temperature gradient:

$$q = - \ k \ \frac{\partial T}{\partial x} \tag{3-9}$$

From eqs. (3-8a) and (3-9) it follows that:

$$\rho c \ \frac{\partial T}{\partial \tau} = k \ \frac{\partial^2 T}{\partial x^2} \tag{3-10}$$

Equation (3-10) is the governing equation of heat conduction within a solid slab of constant properties and in the absence of any heat sinks or sources. The boundary conditions, the initial condition and the solution for two cases of main interest are given below:

A. Semi-Infinite Slab Heated Radiantly at the Surface (x = 0)

At the surface $x = 0$, the conductive heat flux in the slab is equal to the absorbed fraction of the radiative flux impinging on the surface*:

$$q_{x=0} = \tilde{\alpha} \ W_o \tag{3-11}$$

Using eq. (3-9), eq. (3-11) becomes:

$$-k \ \frac{\delta T}{\delta x} \ (x = 0, \ \tau) = \tilde{\alpha} \ W_o \tag{3-11a}$$

Also, for a semi-infinite slab, the temperature approaches

_____

*For simplicity it is assumed that all the energy absorption occurs at the surface, $x = 0$.

117

the ambient temperature as x approaches infinity:

$$T(x \to \infty, \tau) = T_o \qquad (3\text{-}12)$$

Finally, the initial temperature of the slab is the ambient temperature:

$$T(x, \tau = 0) = T_o \qquad (3\text{-}13)$$

From reference [27], the solution of Eq. (3-10) with the conditions (3-11a), (3-12) and (3-13) yields:

$$T(x, \tau) = T_o + \left[\frac{2\, \tilde{\alpha} W_o}{k} \cdot \left(\frac{k\tau}{\rho c}\right)^{1/2} \cdot \text{ierfc}\left(\frac{x}{2\left(\frac{k\tau}{\rho c}\right)^{1/2}}\right) \right] \qquad (3\text{-}14)$$

where $\quad \text{ierfc}(y) \equiv \int_y^{\infty} \text{erfc}(z)\, dz$

$$\text{erfc}(y) \equiv 1 - \text{erf}(y)$$

and $\quad \text{erf}(y) \equiv \dfrac{2}{\sqrt{\pi}} \displaystyle\int_o^y e^{-z^2} dz$, is called the error

function and tabulated in reference [27].

At the surface, $x = 0$, Eq. (3-14) reduces to:

$$T(0, \tau) = T_o + \frac{2\, \tilde{\alpha} W_o}{k} \left(\frac{k\tau}{\rho c \pi}\right)^{1/2} \qquad (3\text{-}15)$$

Rearranging Eq. (3-15) and specifying $T_{i,m}$ at $\tau_{i,m}$, yields:

$$\frac{4}{\pi} \tau_{i,m} \cdot W_o^2 = \frac{\rho c k (T_{i,m} - T_o)^2}{\tilde{\alpha}^2} \qquad (3\text{-}16)$$

118

or

$$(N_{Fo})_{i,m} = \frac{\pi}{4} (q^*_{rad})^{-2} \qquad (3\text{-}16a)$$

Eq. (3-16a) gives the normalized ignition or melting
time as a function of the normalized heating intensity
for a semi-infinite slab·

The solution for the heating response of the thermally
thin slab without heat losses (Eq. (3-2a)) and with losses
(Eq. (3-7a) with the Biot number, $N_{Bi} = 0.7$) and for the
heating response of the front surface of the semi-infinite
slab are plotted for comparison on a log-log graph in
Fig. 3-3. It should be noted that at high normalized
heating rates (q^*_{rad}), the solutions of the thermally thin
slab with and without losses merge, i.e., convective losses
become relatively negligible. Their importance increases,
however, as the heating rates decreases. This is evident
from the sharp rise in heating time as the convective heat
losses approach the heating intensity (at $N_{Bi} \simeq q^*_{rad} \simeq 0.7$).
Clearly, under these conditions all the absorbed energy is
lost and ignition or melting does not occur. It should be
noted that if the relative strength of the convective losses
increases, the sharp increase in heating time will occur
at a higher heating intensity.

B.   Finite Slab of Thickness, δ, Heated Radiantly at One
     Surface $(x = \delta)$

As derived in Section 3.2.2A, it can be shown that the
radiant heat flux at the surface $x = \delta$ (see Fig. 3-4) yields
the condition:

$$k \frac{\partial T}{\partial x} (x = \delta, \tau) = \tilde{\alpha} W_o \qquad (3\text{-}17)$$

119

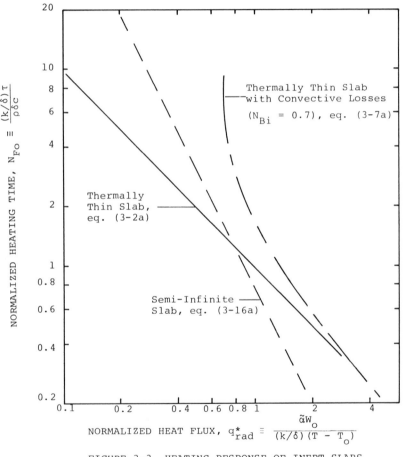

FIGURE 3-3 HEATING RESPONSE OF INERT SLABS

The figure's axis labels and annotations:

NORMALIZED HEATING TIME, $N_{FO} \equiv \dfrac{(k/\delta)\tau}{\rho\delta c}$

NORMALIZED HEAT FLUX, $q^*_{rad} \equiv \dfrac{\tilde{\alpha}W_o}{(k/\delta)(T - T_o)}$

Thermally Thin Slab with Convective Losses (N_{Bi} = 0.7), eq. (3-7a)

Thermally Thin Slab, eq. (3-2a)

Semi-Infinite Slab, eq. (3-16a)

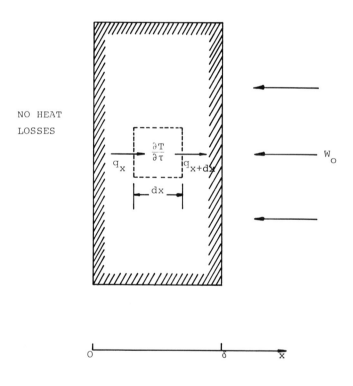

FIGURE 3-4   SLAB OF FINITE THICKNESS, δ

121

Similarly, the assumption of no heat losses at the surface $x = 0$ yields:

$$\frac{\partial T}{\partial x} (x = 0, \tau) = 0 \qquad (3\text{-}18)$$

The initial temperature of the slab is:

$$T(x, \tau = 0) = T_o \qquad (3\text{-}19)$$

From Reference [27], the solution of Eq. (3-10) with conditions (3-17) to (3-19) is:

$$T(x, \tau) = T_o + 2 \frac{\tilde{\alpha} W_o}{k} \left(\frac{k\tau}{\rho c}\right)^{1/2} \sum_{n=0}^{\infty} \left[\text{ierfc} \frac{(2n + 1)\delta - x}{2\left(\frac{k\tau}{\rho c}\right)^{1/2}} \right.$$

$$\left. + \text{ierfc} \frac{(2n + 1)\delta + x}{2\left(\frac{k\tau}{\rho c}\right)^{1/2}} \right] \qquad (3\text{-}20)$$

or

$$2(N_{Fo})^{1/2} \sum_{n=0}^{\infty} \left[\text{ierfc} \frac{(2n+1) - \frac{x}{\delta}}{2\left(N_{Fo}\right)^{1/2}} + \text{ierfc} \frac{(2n+1) + \frac{x}{\delta}}{2\left(N_{Fo}\right)^{1/2}} \right] = (q_{rad}^*)^{-1}$$

$$(3\text{-}20a)$$

Eq. (3-20a) gives the normalized heating time versus the normalized heating intensity at different locations (x/δ) within the finite thickness slab. Eq. (3-20a) is plotted on log-log graph in Fig. 3-5 for two cases:

   (a)   the slab surface exposed to the heat flux, i.e., $x/\delta = 1$

   (b)   the back surface of the slab where $x/\delta = 0$

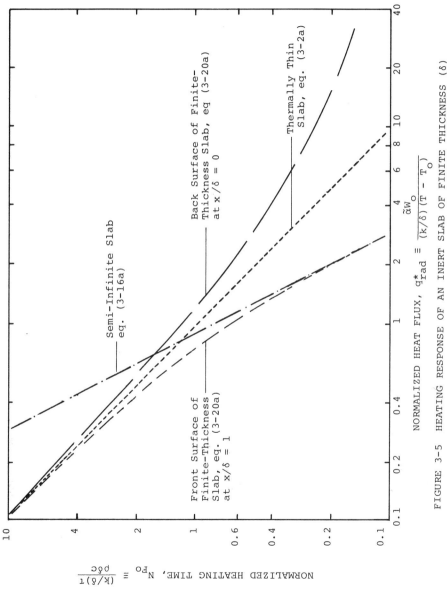

FIGURE 3-5   HEATING RESPONSE OF AN INERT SLAB OF FINITE THICKNESS (δ)

The solutions for the thermally thin and the semi-infinite
cases are also included for comparison.

Notice that for very low normalized heating intensity,
the front and back surfaces of the slab reach a certain
temperature at about the same time and merge with the thermally thin solution. As the normalized heating intensity
increases, the thermal penetration decreases and the front
surface reaches a certain temperature at a much earlier
time than the back surface; temperature gradients develop
within the slab. At $q_{rad}^* \simeq 2$ the front surface solution
merges with that of the semi-infinite slab. Thus, at low
normalized heating intensities, a finite-thickness slab
behaves as if it were thermally thin, while at high normalized heating intensity, it behaves as if it were thermally thick.

3.3 PRE-IGNITION HEAT TRANSFER THROUGH A FABRIC-SKIN SIMULANT SYSTEM

A study was undertaken at FMRC [R9-R13] to assess the
possibility that pain-activated evasive action might reduce the
probability of fabric ignition. Presumably, if a person senses
the presence of a heat source (by feeling pain due to thermal
heating of the skin), he may withdraw from the heat source and
thus prevent ignition of his garment. Experimental and theoretical char (or melt) times were determined for the GIRCFF
fabrics and then compared to the pain times[+] for corresponding
conditions. It was indicated that such pain-activated evasive
action is not likely. Consequently, this phase of the study
was discontinued and replaced by a study of the ease of ignition
of fabrics [R14-R17].

For the sake of completeness, the highlights of the pre-
ignition heat-transfer study are reported below.

[+] Defined (in the manner of Stoll [28]) as the time to reach
a skin temperature of 122°F, see also Chapter 5, Section 5.2.

3.3.1 Experimental Procedure

As illustrated in Fig. 3-6, a fabric-transite-skin-simulant assembly [R12] was exposed to an infrared radiant heat flux of up to about 110,000 BTU/ft^2 hr. (34 W/cm^2). The twenty fabrics provided by the GIRCFF Committee were used and the spacing between the skin-simulant and the fabric was varied from 0 to 1 inch. The temperatures were measured at the unexposed side of the fabric and at the skin-simulant surface, and the time to char or melt was noted.

3.3.2 Experimental and Theoretical Results

The sample of the experimental results given in Table 3-1 shows that pain time is longer than the char (or melt) time. Consequently, pain-activated evasive action (by the wearer) cannot occur prior to ignition. A similar conclusion is also obtained for flaming ignition sources [R13].

A theory was developed at FMRC to study the radiative heating of the three-layer composite system, viz. fabric-air gap-skin. The theory followed essentially the analysis presented in Section 3.2. The fabric was assumed to be inert, i.e. thermal decomposition (or pyrolysis) was neglected. The two cases of dry and moist diathermanous fabric were considered.

To verify the theory, temperature-time histories were measured in the skin-simulant and at the back surface of the fabric, and compared with the theoretical predictions. This is illustrated in Figs. 3-7 and 3-8 for fabrics #4 (100% cotton) and #13 (100% acetate). Because of its location and size, the thermocouple placed in the skin simulant was believed to measure a mean temperature between the depths of 0.5 and 1 mm. As illustrated in Figs. 3-7 and 3-8, its response is between the theoretical predictions for two depths, X = 0.502 mm and X = 1.004 mm, which confirms the theory.

Note in the case of fabric temperature that the

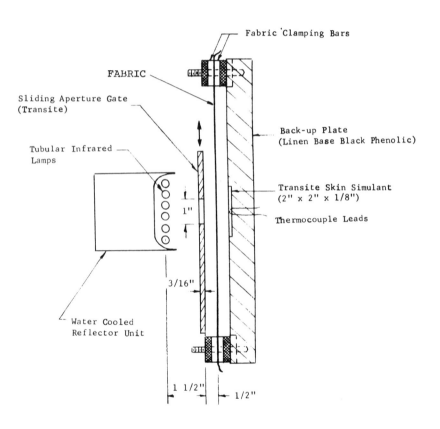

FIGURE 3-6 FABRIC-SKIN (SIMULANT) SYSTEM [R12]

126

TABLE 3-1

EXPERIMENTAL RESULTS [R 12]

FABRIC-SKIN GAP = 1/2"

HEAT SOURCE = 20 BTU/FT^2-SEC

TEMP. ENVIRONMENTAL CHAMBER = 70°F

FABRIC # AND TYPE	REL HUM %	TIME TO CHAR/MELT SEC	TIME TO FLAME SEC	CHAR TEMP. °F	PAIN TIME $\Delta T = 52°F$ SEC	PAIN TIME / CHAR TIME
2 100% P.E.	10	1.4		585	2.4	1.7
	90	1.5		585	2.5	1.7
5 100% COTTON T-SHIRT	10	2.8	3.1	905	4.5	1.6
	90	2.7	5.4	671	5.5	2.0
7 100% ACRYLIC	10	4.5		649	5.2	1.2
	90	1.3		585	5.1	3.9
8 65/35 PE/C	10	5.3	6.5	820	6.6	1.3
	90	2.9	6.0	606	5.8	2.0
10 100% COTTON BATISTE	10	2.5		799	4.6	1.8
	90	2.6	2.9	968	3.5	1.4
11 ACET/NYLON TRICOT	10	2.2		628	4.3	2.0
	90	2.3		692	3.7	1.6
12 100% NYLON TRICOT	10	1.4		628	4.0	2.9
	90	2.2		431	3.8	1.7
13 100% ACET TRICOT	10	2.3		671	4.4	1.9
	90	2.5		649	3.9	1.6
18 100% COTTON FLANNEL	10	2.4	4.2	692	6.0	2.5
	90	4.1	5.4	692	5.9	1.4
19 100% FR COTTON FLANNEL	10	3.0		563	5.4	1.8
	90	3.3		671	9.3	2.8

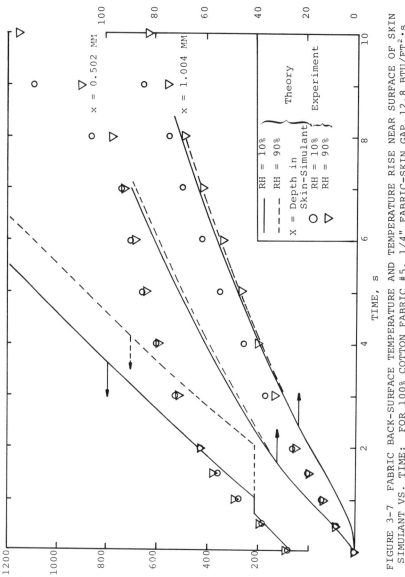

FIGURE 3-7 FABRIC BACK-SURFACE TEMPERATURE AND TEMPERATURE RISE NEAR SURFACE OF SKIN SIMULANT VS. TIME: FOR 100% COTTON FABRIC #5, 1/4" FABRIC-SKIN GAP, 12.8 BTU/FT^2·s [R12]

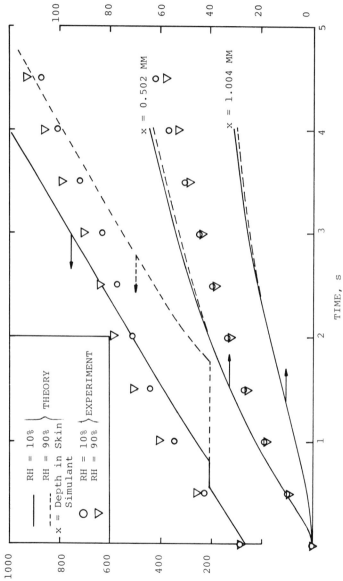

FIGURE 3-8 FABRIC BACK-SURFACE TEMPERATURE AND TEMPERATURE RISE NEAR SURFACE OF
SKIN SIMULANT VS. TIME: FOR 100% ACETATE FABRIC #13, 1/4" FABRIC-SKIN GAP,
12.8 BTU/FT² · s [R12]

effect of relative humidity (from 10% to 90%) is much larger
for the predicted values than for the experimentally measured
values. Agreement between theory and experiment for fabric
temperature is good for fabric #13, but less so for fabric
#5.

3.4 RADIATIVE IGNITION

An extensive radiative ignition study has been undertaken at GIT [R1 to R4, 28a and 28b]. In the following sections, GIT's experimental results are presented and discussed, followed by a consideration of the fabric parameters which may rank the GIRCFF fabrics with respect to "ease of ignition (or melting)" under radiative heating.

3.4.1 Experimental Procedure at G.I.T.

Fabric samples of 1 in. diameter were exposed at G.I.T. to uniform constant heat fluxes of 0.25 to 16 W/cm^2 (800 to 50,000 Btu/hr ft^2). The exposure times varied between 1 and 90 seconds. The experimental apparatus was enclosed in a thermostatically and psychrometrically controlled chamber. Upon opening of a shutter mechanism as shown in Fig. 3-9 the fabric was exposed to a Modular Radiant Heater which used tungsten-filament tubular quartz lamps. In testing cotton and cotton/polyester blend fabrics, an infrascope was used to detect the occurrence of a flame on the backside of the exposed fabric. For thermoplastic fabrics, the fabric was placed on a line between a high-intensity lamp and the infra-scope and when the fabric melted through, the infrascope detected the light beam. The destruction time was defined as the time interval from the opening of the shutter (as sensed by a micro-switch) to the detection of a flame or a melt-through (as sensed by the infrascope). Further details on the experimental apparatus are given in references R1 to R4.

3.4.2 Results

It is known that, at any heating intensity, ignition times will depend on fabric thermophysical properties. Normalizing ignition results with respect to the thermophysical properties in a manner such as that described in Section 3.2, allows for presentation of all data for the twenty fabrics in a single plot. The GIT results are so plotted in Fig. 3-10, which

131

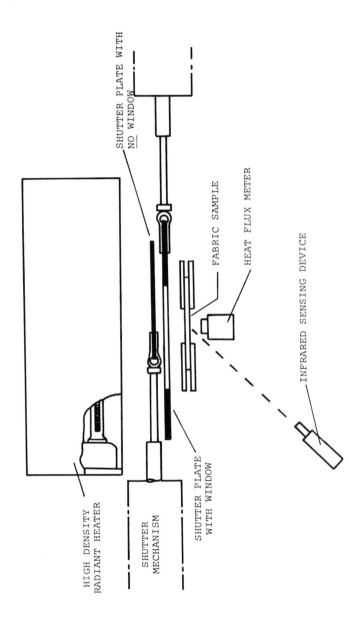

HIGH DENSITY
RADIANT HEATER

SHUTTER
MECHANISM

SHUTTER PLATE
WITH WINDOW

SHUTTER PLATE WITH
NO WINDOW

FABRIC SAMPLE

HEAT FLUX METER

INFRARED SENSING DEVICE

FIGURE 3-9   ARRANGEMENT OF HEATER, SHUTTER, SAMPLE AND SENSORS
IN IGNITION TIME APPARATUS   [R4]

shows the normalized destruction time, $(N_{Fo})_{i,m}$ vs. normalized radiative heating intensity, q^{**}_{rad}, where:

$$(N_{Fo})_{i,m} = \frac{(k/\delta)\tau_{i,m}}{(\rho\delta)c} \tag{3-3}$$

$$q^{**}_{rad} = \frac{(1-\tilde{\rho})W_o}{(k/\delta)(T_{i,m} - T_o)} \tag{3-21}$$

and $\tilde{\rho}$ = reflectivity for charred fabric at $(T_{i,m} + T_o)/2$.

Notice that the normalized radiative intensity eq (3-21) has been defined by GIT in a manner slightly different from that used in Section 3.2, eq. (3-4). Specifically, in the GIT definition of q^{**}_{rad} the term $(1 - \tilde{\rho})$ replaces the absorptivity $(\tilde{\alpha})$ used in eq. (3-4). The Overview Team reasoned that since the fraction of the impinging flux absorbed by the fabric is determined by its absorptivity $(\tilde{\alpha})$ and $\tilde{\alpha} = (1 - \tilde{\rho})$ only for the case when the transmissivity (τ) is equal to zero[+], then it seems logical to use the more general form of q^{*}_{rad} involving $\tilde{\alpha}$, that is eq. (3-4). Accordingly, the GIT experimental normalized ignition (or melting) times have been replotted in Fig. 3-11 on the basis of the more general concept of the normalized heating intensity. In addition, the theoretical heating response for a thermally thin inert slab is plotted also, according to eq. (3-2a), as well as the heating response for a semi-infinite inert slab according to eq. (3-16a).

It is evident from Fig. 3-11 that the measured fabric ignition times are higher than those predicted for an inert slab. This is understandable because the inert model neglects natural convective heat losses and endothermic processes such as pyrolysis, melting and vaporization. These processes would all tend to retard the rise in temperature and thus increase the ignition (or melting) time.

[+]By definition: $\tilde{\alpha} + \tilde{\rho} + \tilde{\tau} = 1$

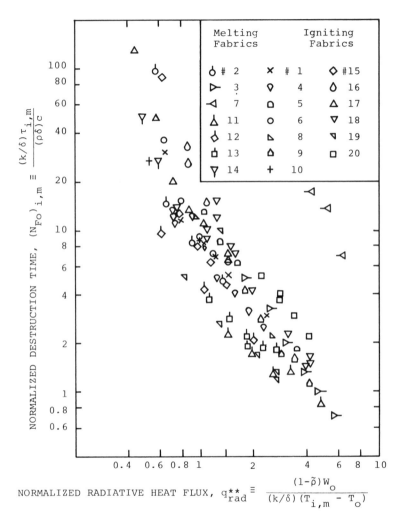

FIGURE 3-10   IGNITION AND MELTING RESPONSE OF GIRCFF FABRICS
FROM WULFF [32]

134

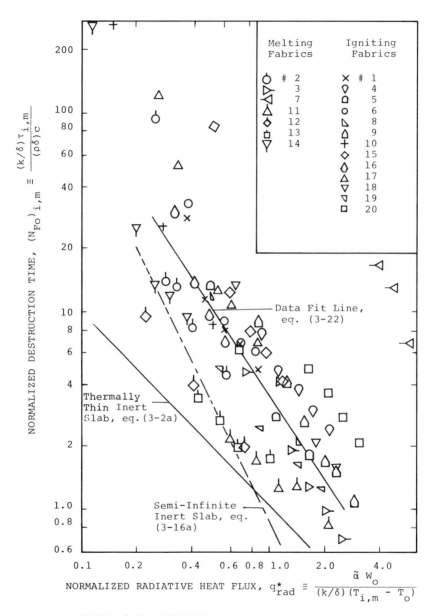

FIGURE 3-11   IGNITION AND MELTING RESPONSE OF GIRCFF
             FABRICS.   DATA ARE TAKEN FROM WULFF [32]

135

It is also evident from Fig. 3-11 that the relative effect of natural convective losses and endothermic processes becomes very important at low heating rates. In fact, ignition may even be prevented at these low heating rates [R4]. If losses are included in the thermally thin inert model, the qualitative dependence of $(N_{Fo})_{i,m}$ on q^*_{rad} is similar to GIT's experimental results. This is illustrated in Fig. 3-3 where N_{Bi} for the heat losses[+] is assumed to be $\simeq 0.7$ (as done by GIT [R4]).

3.4.3 Discussion of Results

The graph of Fig. 3-11 shows some interesting features. For example, it is seen in the ignition tests for cotton and polyester/cotton blends, that ignition time is measured for the occurrence of a flame on the back side and not the exposed side[++] of the fabric (refer to Fig. 3-9). GIT has estimated a negligible difference[+++] between fabric front side and back side temperatures, from the numerical integration of the heat diffusion equation for a thermally thick and finite slab with in-depth radiative energy absorption and convective surface cooling.

It is also worth noting that values of N_{Fo} for melting fabrics (tagged symbols in Fig. 3-11) are consistently lower than those for igniting fabrics, except for the acrylic fabric #7 whose unusual results are discussed below.

[+]The Biot Number determines the strength of the convective heat losses.

[++]In work subsequent to the GIRCFF program, the GIT group did measure ignition time at the front face of the fabric where the additional experimental complexity was justified (as per Private Communication with GIT Group, August 12, 1974).

[+++]Less than 10% except when $q^*_{rad} > 5.0$, cf. Reference R4, p. 78.

This is also true for GIT's convective (i.e. flame) ignition
results (except for fabric #2) as shown later in Section 3.5.2,
Fig. 3-25. As yet, there is no obvious explanation for this
trend.

As can be seen in Fig. 3-11, the ignition results of
fabric #7 are qualitatively the same as those for other fabrics,
but they are obviously quantitatively different (non-dimensional
melting times being one order of magnitude higher than the average results for the other fabrics). Yet, FMRC's results show
no unusual difference between this fabric and the other GIRCFF
fabrics. The FMRC measurement of the melting time for this
fabric under radiative heating (1.3s and 4.5s given in Table 3-1
in Section 3.3.2) and under convective heating (0.35 and 1.6s,
as given in Table 3.5 in Section 3.5.2) is within the range
measured for the other fabrics. Furthermore, an inspection of
the fabric itself shows no unique properties (other than being
the only acrylic fabric) that could cause slow melting. Repeat
tests on acrylic fabric #7 appear desirable.

The definition of q^*_{rad} (eq. 3-4), while physically more
appropriate, has yielded a larger spread of the data as can be
seen by comparing Figs. 3-10 and 3-11. The larger spread
is caused by the correction factor ($\tilde{\alpha}/1 - \tilde{\rho}$) used to calculate
q^*_{rad} in eq. (3-4). As illustrated in Table 3-2, this factor
has a different value for each fabric.

3.4.4 Ranking of GIRCFF Fabrics with Respect to Ease of Ignition (or Melting) under Radiative Heating, as Proposed by Overview Team

In this Section, an attempt is made by the Overview Team
to correlate the ignition (or melting) times reported by
GIT with fabric properties on a semi-empirical basis.
The correlation yields parameters characteristic of the

137

TABLE 3-2

CHARRED FABRIC OPTICAL PROPERTIES USED IN GENERATING FIGURE 3-11

Fabric Number	Reflectivity $\tilde{\rho}$	Absorptivity $\tilde{\alpha}$	Correction Factor $\dfrac{\tilde{\alpha}}{1-\tilde{\rho}}$	Fiber Content
1	.37	.35	.55	65/35 PEC
2	.58	.18	.42	PE
3	.51	.20	.41	PE
4	.40	.51	.85	Cotton
5	.49	.23	.44	Cotton
6	.34	.37	.56	65/35 PEC
7	.11	.72	.82	ACRYLIC
8	.35	.35	.53	65/35 PEC
9	.63	.24	.65	Cotton
10	.22	.39	.50	Cotton
11	.57	.18	.41	78/22 ACET./N
12	.47	.19	.35	Nylon
13	.49	.18	.36	ACETATE
14	.40	.19	.32	Nylon
15	.24	.60	.79	65/35 PER
16	.32	.24	.36	50/50 PEC
17	.26	.43	.57	65/35 PEC
18	.53	.24	.51	Cotton
19	.25	.50	.66	Cotton
20	.11	.61	.68	Wool

fabric thermophysical properties, and it is proposed to use these parameters to rank the GIRCFF fabrics with respect to "ease of ignition" (or melting) under a specified radiative heat flux.

A. Semi-empirical Correlation of Experimental Results

As illustrated in Fig. 3-12, the radiative ignition and melting results for individual fabric (taken from Fig. 3-11) follow the same qualitative trend--regardless of the material. As noted before, the normalized melting time tends to be somewhat lower than the normalized ignition time, but this does not change the general qualitative trend. It is found that the test data for individual fabrics can be represented by simple power relationships:

$$(N_{Fo})_{i,m} \simeq c (q_{rad}^{*})^{n} \tag{3-22}$$

where c and n are determined for each fabric.

This power relationship can be compared with the results of heating an inert slab to a temperature $T_{i,m}$, as derived in Section 3.2 and illustrated in Fig. 3-3:

1. For a thermally thin slab, it has been shown that:

$$(N_{Fo})_{i,m} = (q_{rad}^{*})^{-1} \tag{3-2a}$$

or

$$\tau_{i,m} \cdot W_{o} = \frac{\rho \delta c (T_{i,m} - T_{o})}{\tilde{\alpha}} \equiv A \tag{3-23}$$

2. For a thermally thick slab, it has been shown that:

$$(N_{Fo})_{i,m} = \frac{\pi}{4} (q_{rad}^{*})^{-2} \tag{3-16a}$$

139

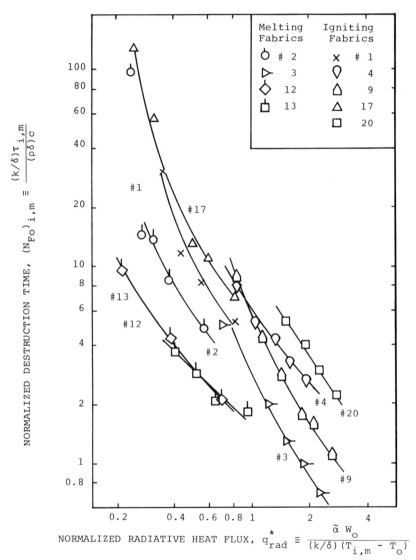

FIGURE 3-12   COMPARISON OF IGNITION AND MELTING RESPONSE FOR
INDIVIDUAL FABRICS.   DATA ARE TAKEN FROM WULFF [32]

140

or

$$\frac{4}{\pi} \cdot \tau_{i,m} \cdot W_o^2 = \frac{(\rho ck)\left(T_{i,m} - T_o\right)^2}{\tilde{\alpha}^2} \equiv B \quad (3\text{-}24)$$

3. And for a thermally thin slab with Newtonian cooling it has been shown that:

$$\left(N_{Fo}\right)_{i,m} = -\frac{1}{N_{Bi}} \ln\left(1 - \frac{N_{Bi}}{q^*_{rad}}\right) \quad (3\text{-}7a)$$

where $N_{Bi} \equiv \frac{h}{k/\delta} \equiv$ Biot Number

and $h \equiv$ coefficient of convective heat transfer, $W/cm^2 \cdot °C$

Given that the case of practical importance is for high heating rates, (as reflected in a high level of q^*_{rad}) where Newtonian-cooling effects are negligible, the experimental results of Fig. 3-11 can be fitted on the average by a straight line in the form of Eq. (3-22) with all the data within about 100% of the average line. This yields $n = -3/2$ and $c = 4$ for $q^*_{rad} > 0.3$. (These values are determined very crudely by a visual best fit of all the data, excluding only those for the acrylic fabric #7. Better values for n and c can be determined and are discussed at the end of this Section.)

Thus, for $q^*_{rad} > 0.3$, Eq. (3-22) becomes:

$$\left(N_{Fo}\right)_{i,m} \simeq 4 \left(q^*_{rad}\right)^{-3/2} \quad (3\text{-}25)$$

or in its more explicit form:

$$\frac{(k/\delta)\,\tau_{i,m}}{(\rho\delta)c} \simeq 4\left[\frac{k/\delta\left(T_{i,m} - T_o\right)}{\tilde{\alpha}W_o}\right]^{3/2} \quad (3\text{-}25a)$$

141

and rearranging the terms of (3-25a) it follows that:

$$\frac{1}{4}\tau_{i,m} \cdot W_o^{3/2} \simeq \frac{(c\rho\delta)(T_{i,m} - T_o)^{3/2}}{\tilde{\alpha}^{3/2}} \left(\frac{k}{\delta}\right)^{1/2} \equiv D \qquad (3\text{-}25b)$$

From eq. (3-25b) it is clear that for a specified radiant heat flux, W_o, the ignition of the 19 GIRCFF fabrics (excluding #7) is determined on the average by the thermal and physical properties of the fabrics (as expressed in the parameter D)-- regardless of their pyrolysis or melting characteristics. (Of course, the individual fabric ignition or melting results depend somewhat on these characteristics.) Furthermore, the parameter D specifies the weighted value of each of these properties. Thus, because of the difference in magnitude of the exponents 3/2, 1 and 1/2, the importance of fabric properties to ignition or to melting is ranked in the following order:

1. The ignition or melting temperature, $T_{i,m}$ and absorptivity, $\tilde{\alpha}$.
2. The thermal capacity/unit area, i.e. $\rho\delta c$.
3. The thermal conductance, k/δ.

This result seems plausible. Simply by its definition, $T_{i,m}$ is expected to be important to the ignition or melting of fabrics. Also $\tilde{\alpha}$ determines the fraction of incident energy flux absorbed by the fabric and it should be most important. On the other hand, because of small fabric thicknesses (and consequently small subsurface temperature gradients) thermal conductance is less important to the ignition process for fabrics than are the other two properties.

It should be noted that the absorptivity of a fabric is a function of the radiation wavelength of the heating source. Thus, the parameter D depends also on the heating source radiation wavelength.

142

For the data correlation presented above, it is important to remember that the exponent n = - 3/2 has been determined from Fig. 3-11 by a very crude way (a visual best fit of the data). A linear regression analysis can, of course, be done to determine the exponents for each variable in eq. (3-25a) to fit best the experimental data (with respect to a criterion such as minimum mean square error). Hallman et al[30] have performed such a mean square fit of ignition data for polymer materials and their results are given in Fig. 3-13. It is interesting to note the resemblance between these results and eq. (3-24) for a heated, thermally thick inert slab (thickness of samples used in Fig. 3-13 $\approx$ 1.27 cm). Note that the abscissa of Fig. 3-13 plots the inverse of the heating intensity.

B. Tentative Ranking of Fabrics

From the semi-empirical correlation presented by the Overview Team in the previous Section (eq. 3-25b), it is clear that for a specified heat flux, the larger the value of the parameter D, the longer the ignition time. Thus, D is tentatively proposed by the Overview Team as a parameter that could rank the 19 GIRCFF fabrics (#7 excepted)* with respect to their "ease of ignition (or melting)" under radiative heating. However, it is important to keep in mind the following limitations:

>1. The parameter D is concerned only with the
>ignition and melting characteristics of the
>fabric. Other considerations, such as the
>probabilities of flame spreading, extinguishment, and
>thermal injury, have to be considered to assess the
>overall hazards associated with a certain fabric.

---

*Pending further testing of this fabric, as discussed in Section 3.4.3.

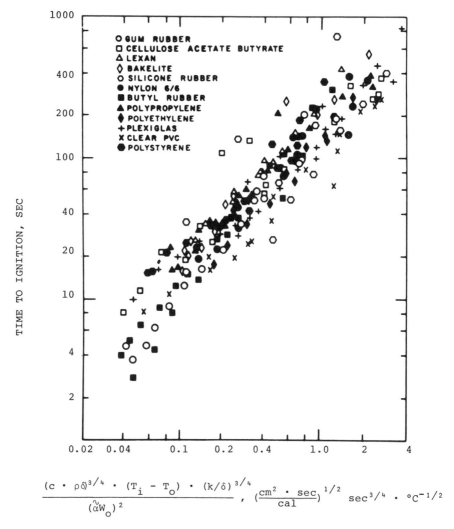

$$\frac{(c \cdot \rho\delta)^{3/4} \cdot (T_i - T_o) \cdot (k/\delta)^{3/4}}{(\tilde{\alpha}W_o)^2} , \ (\frac{cm^2 \cdot sec}{cal})^{1/2} \ sec^{3/4} \cdot {}^{\circ}C^{-1/2}$$

FIGURE 3-13 GENERALIZED CORRELATION OF POLYMER IGNITION DATA
CALCULATED FROM A MATHEMATICAL MODEL.   THE PLOT SHOWS A CLOSE
CORRELATION FOR A LARGE NUMBER OF MATERIALS DESPITE THE FACT
THEIR THERMAL PROPERTIES AND IGNITION TEMPERATURES WERE NOT
DETERMINED FIRST HAND, BUT OBTAINED FROM THE LITERATURE [30]

2. The parameter D is determined on the average.
Thus, depending on the individual pyrolysis or
melting characteristics of a fabric, the actual
ignition time will be higher or lower than that
estimated using D. This is evident from Figs. 3-11
and 3-12. Hence, the parameter D should be considered
only as a rough estimate. Future work could be
devoted to determine a modified D-parameter that
includes the pyrolysis and melting characteristics
of fabrics in addition to their thermophysical
properties. It is expected that this modified
D-parameter would provide improved ranking of fabrics
in relation to "ease of ignition or melting".

To remedy the limitation on D described in the
previous paragraph, an alternative parameter is
presented below. From Fig. 3-11 it is seen that the
prediction of $(N_{Fo})_{i,m}$ by an inert heating model for
a thermally thin slab is <u>always smaller</u> than the
experimental results for the fabrics. Hence, the
<u>thermally thin inert slab</u> model represents a <u>lower
bound</u> for the ignition or melting results for
fabrics. It is very desirable from ignition safety
considerations to identify such a lower bound (and
then to determine it if is acceptable). From
eqs. (3-2) and (3-23) this lower bound can be des-
ignated by the parameter

$$A \equiv \frac{\rho \delta c (T_{i,m} - T_o)}{\tilde{\alpha}} \qquad (3-23)$$

145

It is obvious that, except for the thermal conductivity, the two parameters A and D contain the
same physical and thermal properties--though weighted
somewhat differently (because of the different
exponents). Thus, some correlation between A and D
is expected. This is shown in Fig. 3-14. Thus, very
roughly speaking, a fabric that ranks high or low
using D will rank in about the same way using A.

In Fig. 3-15, the parameter A is correlated to
the mass/unit area, $\rho\delta$--a property well known to the
textile industry. As expected, the correlation is
not very good because fabrics at the same mass/unit
area may vary significantly in their thermal properties. It is still true, however, that (other
thermal properties being equal) the heavier the
fabric, the slower it will ignite (or melt).
Figure 3-16 shows that the correlation between D
and $\rho\delta$ is even worse (than for A and $\rho\delta$) because
the thermal properties are weighted even more in
the calculation of D.

Tables 3-3 and 3-4 rank the GIRCFF fabrics with
respect to their lower bound on the ignition and
melting times, respectively. Because ignition and
melting have different hazard implications (as
discussed in Section 3.1), it is important to keep the
A parameters calculated for ignition separate from
those calculated for melting. Tables 3-3 and 3-4
show that there is no consistent trend in the
ranking due to the fiber content of the fabric. The
range in the A value is about three-fold, but the
change in A from one fabric to another is small.

To modify the ignition ranking order of any
fabric, one or more of its thermophysical properties
which comprise the parameter A can be modified.

146

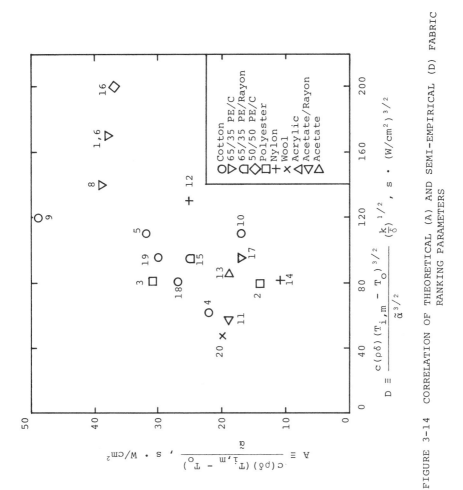

FIGURE 3-14 CORRELATION OF THEORETICAL (A) AND SEMI-EMPIRICAL (D) FABRIC
RANKING PARAMETERS

147

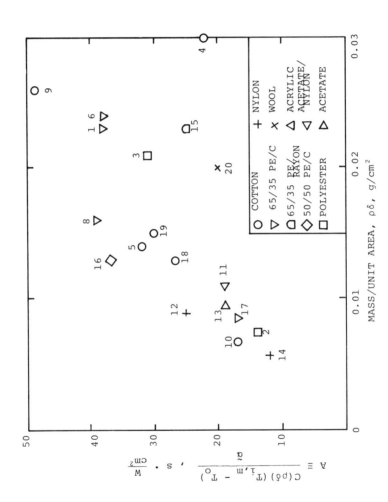

FIGURE 3-15 FABRIC RANKING PARAMETER (A) VS. MASS/UNIT AREA ($\rho\delta$)

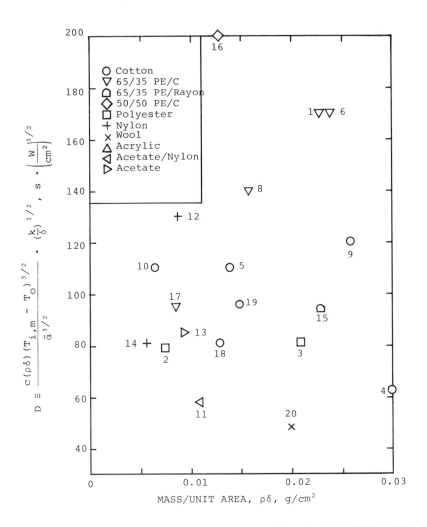

FIGURE 3-16 FABRIC RANKING PARAMETER (D) VS. MASS/UNIT AREA (ρδ)

149

TABLE 3-3

FABRIC RANKING WITH RESPECT TO LOWER BOUND ON IGNITION TIMES
UNDER RADIATIVE HEATING, AS PROPOSED BY OVERVIEW TEAM

Ranking Order	$A \equiv \rho\delta c(T_i - T_o)/\tilde{\alpha}$ $W \cdot s/cm^2$	GIRCFF Designation Number	Fiber Content
1	49	9	Cotton
2	39	8	65/35 PEC
3	38	1	65/35 PEC
		6	65/35 PEC
5	37	16	50/50 PEC
6	32	5	Cotton
7	30	19	Cotton, FR
8	27	18	Cotton
9	25	15	65/35 PE/Rayon
10	22	4	Cotton
11	20	20	Wool
12	17	10	Cotton
		17	65/35 PEC

TABLE 3-4

FABRIC RANKING WITH RESPECT TO LOWER BOUND ON MELTING TIMES
UNDER RADIATIVE HEATING, AS PROPOSED BY OVERVIEW TEAM

Ranking Order	$A \equiv \rho\delta c(T_m - T_o)/\tilde{\alpha}$, $W \cdot s/cm^2$	GIRCFF Designation Number	Fiber Content
1	31	3	PE
2	25	12	Nylon
3	19	11	Acetate
		13	78/22 Acetate/Nylon
5	14	2	PE
6	12	14	Nylon

A quick review of eq. (3-23) shows the physical logic of this claim.

3.4.5 Summary of Results of Ranking

The main conclusions summarized below are applicable for radiative heating only:

1. A semi-empirical correlation of the ignition (or melting) times and the thermophysical properties obtained in this study yields:

$$\frac{1}{4} \cdot \tau_{i,m} \cdot W_o^{3/2} = \frac{(\rho\delta c)\,(T_{i,m} - T_o)^{3/2}}{\tilde{\alpha}^{3/2}} \left(\frac{k}{\delta}\right)^{1/2} \equiv D \qquad (3\text{-}25b)$$

    Thus, at any heat flux, the ignition (or melting) times, $\tau_{i,m}$, can be expressed in terms of the fabric parameter D.

2. The parameter D has possible utility as a guideline for modification of the ignition (or melting) behavior of fabrics.

3. A fabric-parameter, $A \equiv \rho\delta c(T_{i,m} - T_o)/\alpha$ (derived from a heat-transfer analysis of an inert thermally-thin slab model), is further proposed as a basis for ranking fabrics with respect to the lower bound of their ignition (or melting) times.

3.5 FLAMING IGNITION

Fabric flaming ignition studies were also undertaken at FMRC[R14-R17], at GIT [R5-R8, 30a] and at GRI[R27]. In the following sections, their experimental procedures and results are presented and compared. A fabric parameter, intended for ranking the GIRCFF fabrics with respect to "ease of ignition (or melting)" under flaming ignition, is then discussed.

3.5.1 Experimental Procedures

In FMRC's experimental program, a 3" x 3" fabric sample, placed at arbitrary orientations in the gravity field, was exposed to diffusion flames (from methane burners and ordinary book matches) and air-methane pre-mixed flames. Burner diameter, fabric orientation, flow rate and air-fuel ratio were controlled, as illustrated in the line diagram of the apparatus in Fig. 3-17. Exposure times were increased by 0.01 sec increments until the "minimum response time" which denoted the onset of self-sustained ignition for fabrics that ignited as the first response or the destruction of one yarn for fabrics that melted as the first response. Additional details on the experimental procedure can be found in reference R17.

In GIT's experimental program two tests were presented, viz., a static test (already in use), where the fabric was exposed to an established flame when a shutter was opened--as illustrated in Figs. 3-18 and 3-19, and a dynamic test (planned for future work), where the fabric was moved at pre-determined speeds in and out of an established flame, as illustrated in Fig. 3-20. For both tests, the fabric sample size was 2.5" diameter and the burner inner diameter was 1.5". A few tests, in the static mode, were carried out with 1" fabric samples. The distance between burner and fabric, the fuel flow rate and the air/fuel ratio were controlled variables of the test. Destruction times were measured by an infrared detector from the onset of flame exposure to the occurrence of ignition (not necessarily self-sustained) or melt-through of the fabric. As illustrated in Fig. 3-21, the infraredscope detected ignition on

152

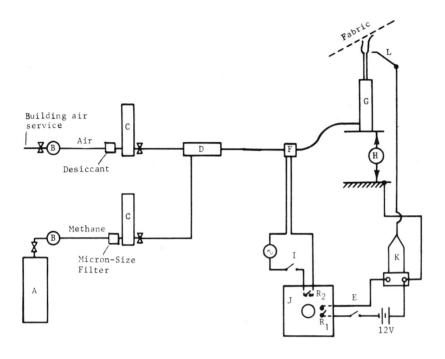

A   METHANE CYLINDER

B   PRESSURE REGULATOR AND GAGES (REDUCED PRESSURE = 10 PSIG)

C   FLOW METERS

D   MIXER

E   SWITCH TO ENERGIZE IGNITION COIL (K)

F   SOLENOID VALVE, NC

G   BURNER WITH INTERCHANGEABLE BURNER TUBES (5 DIAMETERS,
    0.030" - 0.50")

H   DIAL GAGE

I   SWITCH TO INITIATE GAS FLOW

J   DIGITAL TIME CONTROL (EAGLE SIGNAL, MODEL CT530)
    (R_1 deactivates at start of time interval and R_2 deactivates at
    end of time interval)

K   AUTOMOBILE IGNITION COIL

L   SPARK ELECTRODE

FIGURE 3-17   SCHEMATIC OF IGNITION APPARATUS   [R17]

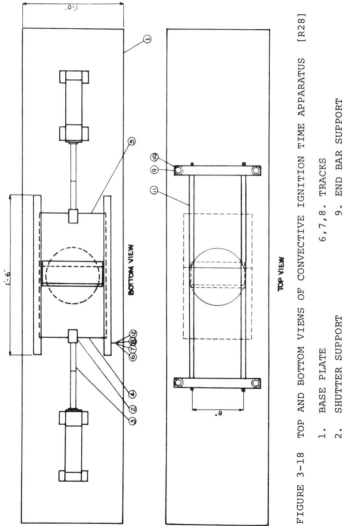

BOTTOM VIEW

TOP VIEW

FIGURE 3-18   TOP AND BOTTOM VIEWS OF CONVECTIVE IGNITION TIME APPARATUS   [R28]

1. BASE PLATE   6,7,8. TRACKS
2. SHUTTER SUPPORT   9. END BAR SUPPORT
3. SUPPORT ROD   10. SUPPORT TUBE
4. SHUTTER PLATE 1   11. SUPERSTRUCTURE ROD
5. SHUTTER PLATE 2   12. TRACK EXTENSION

154

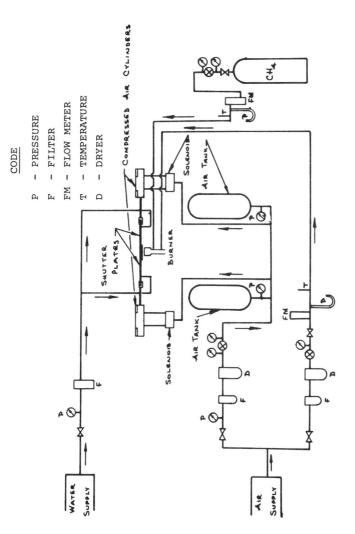

CODE

P – PRESSURE
F – FILTER
FM – FLOW METER
T – TEMPERATURE
D – DRYER

FIGURE 3-19   SCHEMATIC OF CONVECTIVE IGNITION TIME APPARATUS WITH
AIR, FUEL AND WATER FLOW PATHS   [R8]

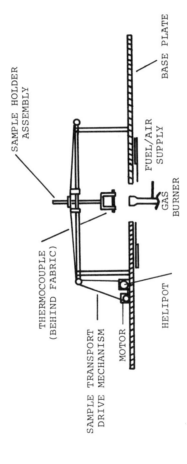

FIGURE 3-20   SCHEMATIC OF CONVECTIVE IGNITION TIME APPARATUS
WITH TRANSPORT SYSTEM [R8]

the exposed side of the fabric (and not on the back side as for
radiative ignition, see Section 3.4.1). Further details on
the experimental procedures can be found in References [R5-R8].

A comparison of FMRC's and GIT's experimental procedures
indicates some differences whose significances are not fully
understood:

1. The relative sizes of fabric to burner were different
for the two procedures, as illustrated in Fig. 3-22. For
FMRC, the flame appeared essentially as a point source in
an infinitely large sample. For GIT, fabric and burner
were approximately of the same size.

2. FMRC measured ignition times only for self-sustained
ignition. However, in the GIT tests, ignition was not
necessarily self-sustained after the removal of the burner
flame. This explains why GIT reported an ignition time
for the Flame Retardant Cotton #19, while FMRC did not,
for this fabric ignites, but does not sustain flaming.
It is clear that extensive, non-localized, burn damage
occurs only if flame spreading follows ignition. Thus,
in hazards assessment, the study of self-sustained
ignition is more meaningful than that of a transient
ignition leading to extinguishment.

Finally, GRI's experimental program[R27] attempted to
simulate a real-life ignition accident, namely, exposure over a
kitchen gas range. The specimens were fabric loops
(9" warp x 22" filling) with a hemmed edge turned back one
inch, sewn and pressed flat. The sleeve-like specimens were
moved horizontally over a gas stove at a fixed speed, and
then withdrawn at the same speed after a specified time interval
of exposure. A schematic diagram of the specimen and burner
is given in Fig. 3-23. The fabric location above the stove
varied from 0.25 to 3 in. The effects of absorbed moisture and
the kind of sewing thread (cellulosic/synthetic) were also
investigated.

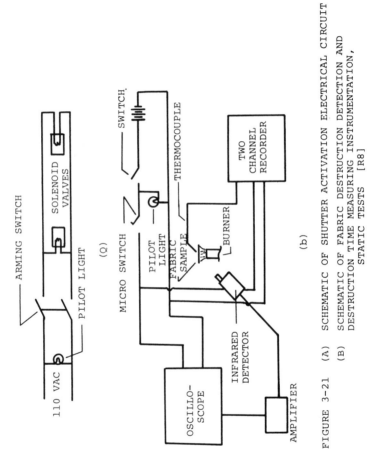

ARMING SWITCH

SOLENOID VALVES

PILOT LIGHT

(a)

110 VAC

SWITCH

MICRO SWITCH

THERMOCOUPLE

PILOT LIGHT

FABRIC SAMPLE

BURNER

TWO CHANNEL RECORDER

OSCILLO-SCOPE

INFRARED DETECTOR

AMPLIFIER

(b)

FIGURE 3-21   (A)   SCHEMATIC OF SHUTTER ACTIVATION ELECTRICAL CIRCUIT
            (B)   SCHEMATIC OF FABRIC DESTRUCTION DETECTION AND
                  DESTRUCTION TIME MEASURING INSTRUMENTATION,
                  STATIC TESTS   [R8]

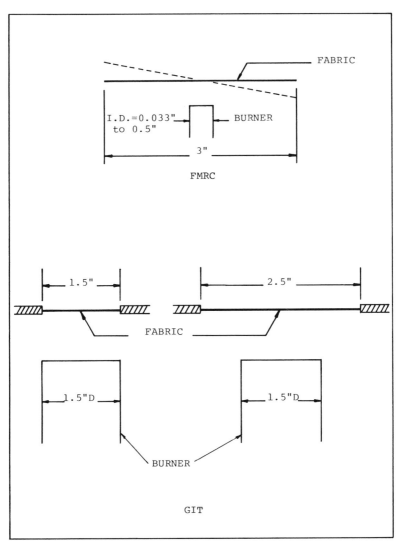

FIGURE 3-22   A COMPARISON OF FMRC's AND GIT's EXPERIMENTAL METHODS

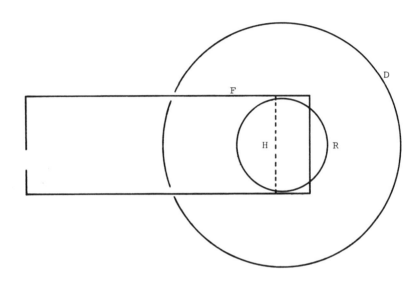

FIGURE 3-23   PROJECTION OF SPECIMEN LOOP ON PLANE OF
BURNER, SCALE 0.5" FOR 1" [R 27]

F = FABRIC

H = HEM

R = RING OF BURNER HOLES

D = DISH OR DEPRESSION IN STOVE TOP,
SURROUNDING THE BURNER

3.5.2 Results and Discussion

FMRC's results have been summarized as follows[R17]:

1. "The response time (ignition time or melting time) of a fabric had a clearly defined minimum with respect to the location of the fabric in the flame, always occurring for fabric locations close to the luminous tip of diffusion flames or the tip of the inner cone for premixed flames."

2. "Minimum response times (as defined above) were insensitive to variations in fabric orientation from the horizontal up to 62° with the horizontal, with two notable exceptions. The exceptions were associated with two highly surface-structured fabrics, a knit-brush and a terry cloth, which had considerably longer ignition times in the horizontal position than in inclined positions."

3. "Minimum ignition times were insensitive to certain variations in burner-tube diameter and flame height for both diffusion and premixed flames. The match flame, which had a flame height in the range considered, yielded minimum ignition times comparable to the results for the methane diffusion flames."

4. "For premixed flames, the air-methane mixture ratio had an important effect on ignition behavior. Minimum ignition times decreased monotonically with increasing mixture ratio until the mixture for maximum burning velocity (minimum flame height at given fuel rate) was reached, then remained approximately constant to mixture ratios approaching stoichiometric."

5. "Experiments with a skin-simulant in close proximity to the unexposed side of the fabric revealed only slight effects on the minimum ignition time."

161

6. "Controlled experiments on effects of the ambient relative humidity indicated a slight decrease in minimum response time with increasing relative humidity only for fabrics of high moisture regain, although RH was varied from 10% to 90%."

7. "From experimental results and a partial theory of fabric ignition under convective heating, it was concluded that fabric porosity is not important." It is interesting to note that Moussa et al[31] arrived at a similar conclusion in flame-spreading studies over cellulosic materials. Figure 3-24 illustrates that for the porosity ranges studied, flame spreading speed depends on the mass/unit area regardless of porosity (as measured by air permeability).

8. The minimum ignition and melting times of the GIRCFF fabrics and their testing conditions are given in Table 3-5.

GIT's results are illustrated in Fig. 3-25 as normalized destruction times, $(N_{FO})_{i,m}$, vs. normalized convective heat flux, q_c^*, where

$$q_c^* \equiv \frac{2h_c(T_f - \overline{T})}{(k/\delta)(T_{i,m} - T_o)} \tag{3-26}$$

h_c = coefficient of heat transfer from flame to fabric as measured by GIT, $W/cm^2 \cdot °C$[†]

[†]Note that eq. (3-26) can be approximated by eq. (3-6), presented in Section 3.2.1, where:

$$T_{avg} \simeq \overline{T} \equiv \frac{1}{\tau_{i,m}} \int_0^{\tau_{i,m}} T \, d\tau$$

and $h_f \equiv 2 h_c$, the appearance of the multiplier 2 is due to the definition of the heat transfer coefficients.

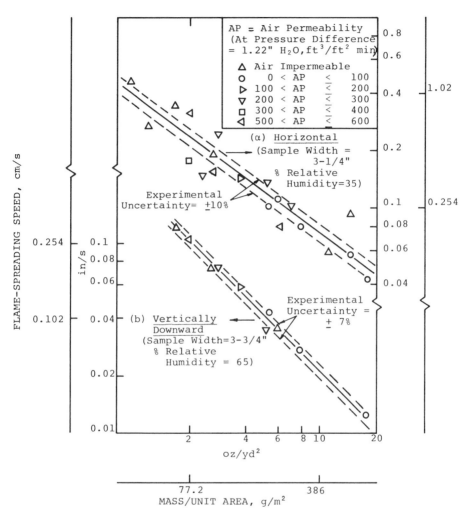

FIGURE 3-24 FLAME-SPREADING SPEED VS. MASS/UNIT AREA FOR
CELLULOSIC MATERIALS OF DIFFERENT AIR PERMEABILITIES. (RIGHT
AND LEFT ORDINATES ARE FOR HORIZONTAL AND FOR VERTICALLY DOWN-
WARD ORIENTATION, RESPECTIVELY.) [31]

TABLE 3-5 MINIMUM IGNITION AND MELTING TIMES FOR GIRCFF FABRICS [R17]

Burner Tube #4; 136 cc/min Methane; Air-Methane Mixture Ratios of 0 (Diffusion Flames) and 7.7 (Premixed Flames). Face* of Fabric Exposed, With Fill Horizontal.

GIRCFF Fabric #	Fabric wt(oz/yd²)	Premixed Flames Melt.time(sec) 0°	41°	Ign.time(sec)** 0°	41°	Premixed Flames Melt.time(sec) 0°	41°	Ign.time(sec)** 0°	41°
1	7.07	-	-	1.2 -1.4	1.2 -1.4	-	-	4.4-4.6	4.2-4.4
2	2.19	0.10-0.11	0.10-0.12	None	0.40-0.45	0.55-0.60	0.45-0.50	?	0.6-0.8
3	5.83	0.35-0.40	0.50-0.55	?	None	1.3 -1.4	1.3 -1.4	?	4.8-5.0
4	8.66	-	-	1.8 -2.0	1.8 -2.0	-	-	5.2-5.4	5.0-5.2
5	3.97	-	-	0.8 -0.9	1.0 -1.1	-	-	2.7-2.9	2.3-2.4
6	6.99	-	-	1.0 -1.2	1.3 -1.5	-	-	4.6-4.8	3.8-4.0
7	4.45	0.35-0.40	0.40-0.45	?	0.9 -1.0	1.5 -1.6	1.3 -1.4	?	1.4-1.6
8	4.82	-	-	1.0 -1.2	0.7 -0.8	-	-	3.4-3.6	2.3-2.5
9	7.42	-	-	3.0 -3.2	0.9 -1.0	-	-	4.7-4.9	1.8-2.0
10	1.90	-	-	0.5 -0.6	0.5 -0.6	-	-	1.4-1.5	1.3-1.4
11	3.00	0.13-0.15	0.19-0.20	?	0.4 -0.5	0.9 -1.0	0.8 -0.9	?	1.2-1.4
12	2.47	0.15-0.16	0.18-0.20	?	None	0.75-0.80	0.7 -0.8	?	1.6-1.9
13	2.17	0.11-0.12	0.08-0.10	?	0.4 -0.5	0.55-0.60	0.50-0.55	?	0.8-0.9
14	1.67	0.10-0.11	0.14-0.15	?	None	0.6 -0.7	0.6 -0.7	?	None
15	6.62	-	-	1.0 -1.1	1.4 -1.5	-	-	3.6-3.8	3.2-3.5
16	3.91	-	-	0.3 -1.0	0.8 -0.9	-	-	2.2-2.4	2.2-2.4
17	2.53	-	-	0.40-0.45	0.45-0.50	-	-	1.5-1.6	1.4-1.5
18	3.69	-	-	0.8 -1.0	0.8 -1.0	-	-	2.2-2.4	2.2-2.4
19	4.21	-	-	None	None	-	-	None	None
20	6.48	1.3 -1.4	1.6 -1.7	?	4.2 -4.4	3.6 -4.0	3.6 -4.0	?	4.8-5.0

*Side of fabric judged commonly exposed in normal wear. To avoid ambiguities, the following clarifications are offered: 1) Twill fabrics were exposed on side with predominant twilling, except for Fabric #4 (denim); 2) Plain/napped fabrics (#18 and #19) were exposed on unnapped side; 3) the double knit (#3) was exposed on side opposite to side structured like the face of a single knit.

**Ignition times for melting fabrics were obtained at fabric separations from burner corresponding to minimum melting times.

$\overline{T}$ = average fabric temperature

$$= \frac{1}{\tau_{i,m}} \int_0^{\tau_{i,m}} T \, d\tau$$

$\tau_{i,m}$ = time to ignition or melting, sec.

and the subscript f indicates flame temperature.
The range of radiative ignition data and the response of an inert
thermally thin slab (eq. 3-5a, derived in Section 3.2.1) are
also included in Fig. 3-25 for comparison. It should be noted
that the non-dimensional data for radiative and convective
ignition fall roughly in the same band. (This comparison will
be discussed further in Section 3.6.) Furthermore, all the
data fall above the approximate solution for the thermally thin
inert slab. As explained for the radiative results (in Section
3.4.2) this could be due to endothermic processes such as
pyrolysis, melting and vaporization. These processes would all
tend to retard the rise in temperature and thus increase the
ignition (or melting) time.

A close inspection of the convective ignition results for
individual fabrics (see also Figs. 3-30 and 3-31), indicates
that the relationship of $(N_{Fo})_i$ vs. q_C^* becomes somewhat concave
downward at low q_C^*. As yet, this effect is not understood.

A comparison of FMRC's and GIT's ignition times vs.
distance of fabric from burner exit is given in Fig. 3-26 for
fabric #10. The results are compatible despite different
testing conditions such as sample size, burner size, fuel flow
rate, and air fuel ratio. The testing conditions are specified
in Fig. 3-26. The higher ignition times measured by GIT
indicate that GIT's testing conditions were far from being the
most stringent (i.e. far from the minimum ignition time location).

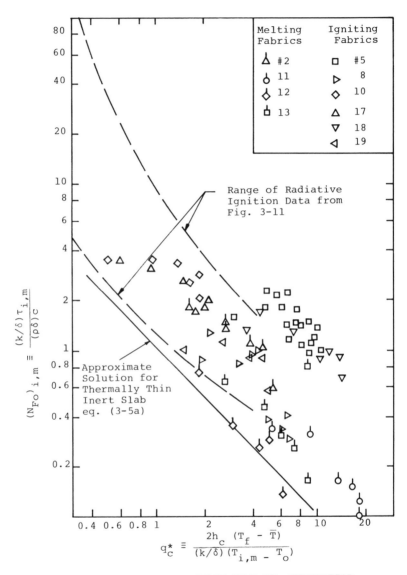

FIGURE 3-25   NORMALIZED DESTRUCTION TIME VS. NORMALIZED
CONVECTION HEAT FLUX [R8]

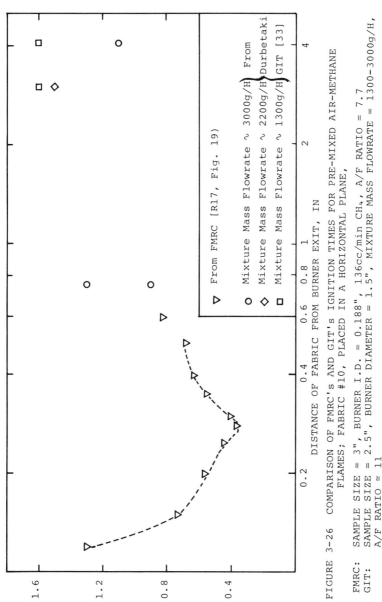

FIGURE 3-26 COMPARISON OF FMRC's AND GIT's IGNITION TIMES FOR PRE-MIXED AIR-METHANE
FLAMES; FABRIC #10, PLACED IN A HORIZONTAL PLANE,

FMRC: SAMPLE SIZE = 3", BURNER I.D. = 0.188", 136cc/min CH₄, A/F RATIO = 7.7
GIT: SAMPLE SIZE = 2.5", BURNER DIAMETER = 1.5", MIXTURE MASS FLOWRATE = 1300-3000g/H,
       A/F RATIO ≈ 11

GRI's results, given in Fig. 3-27, indicate the importance of weight for determining the average time to ignition for fabrics containing cellulose. They also reported the probability of ignition of the GIRCFF fabrics placed at 1" above stove for exposure times of 5, 10 and 15 sec. (20 specimens have been tested for each fabric, as shown in Table 3-6[R-27]). Also, the effect of height above the stove is illustrated in Fig. 3-28. This figure indicates that the higher the sleeve from the stove, the lower the probability of ignition for any specified exposure time; or, in other words, the higher the sleeve from the stove, the greater the ignition time. This result is in agreement with FMRC's results (as given in Fig. 3-26) for a fabric-flame spacing $\gtrsim$ 0.3". It is unfortunate that GRI did not use spacings smaller than 0.5" in its tests. This would have determined the minimum ignition time of sleeve-like specimens.

3.5.3 Tentative Ranking of GIRCFF Fabrics with Respect to Ignition (or Melting) Times under Convective Heating, as Proposed by Overview Team

As illustrated in Fig. 3-25, the ignition and melting times under convective heating have been determined for only ten GIRCFF fabrics by GIT. Thus, a semi-empirical correlation (similar to that performed for the radiative heating results in Section 3.4.4) is not attempted. When more experimental data are available, it will be fruitful to determine (for convective heating) a parameter corresponding to D. This parameter will guide the textile designer in modifying fabric behavior under this heating mode.

It should be noted, however, that the experimental results (in Fig. 3-25) are always higher than those predicted for the heating of an inert, thermally thin slab model. Assuming that this would also be true for the ten remaining (non-tested) GIRCFF fabrics[†], this model can be used to give a lower bound for the

---

[†]While this assumption seems plausible, it should be verified by testing the ten remaining GIRCFF fabrics. If their behavior comes out differently, then the results in this section are applicable only to the ten GIRCFF fabrics tested.

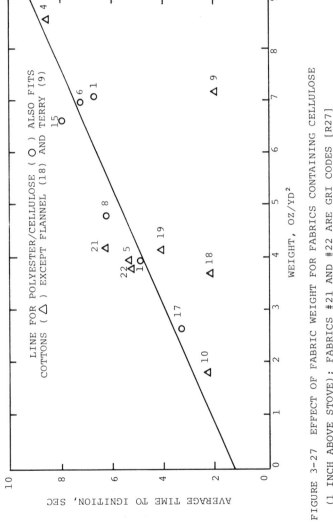

FIGURE 3-27 EFFECT OF FABRIC WEIGHT FOR FABRICS CONTAINING CELLULOSE

(1 INCH ABOVE STOVE); FABRICS #21 AND #22 ARE GRI CODES [R27]

TABLE 3-6

PROBABILITY OF IGNITION AT 5, 10, or 15

SECOND EXPOSURE [R27]

Fabric No.	Fiber	Fabric Wt. oz/sq yd	Probability of Ignition 5 sec. %	10 sec. %	15 sec. %	No. Ignited
A. COTTON						
10 Batiste	100% Cotton	1.90	100	100	100	20
18 Flannel	100% Cotton Napped	3.69	90	100	100	20
22 Broadcloth	100% Cotton	3.80	45	90	100	20
5 T-Shirt Jersey	100% Cotton	3.97	60	85	100	20
21 Broadcloth	100% Cotton	4.10	35	90	100	20
19 Flannel	100% Cotton Napped Fire R.T.D.	4.21	85	90	95	20
9 Terry	100% Cotton	7.42	100	100	100	20
4 Denim	100% Cotton	8.66	< 5	90	100	20
B. POLYESTER/CELLULOSE						
17 Batiste	65/35% PE/ Cotton	2.53	80	95	100	20
16 Shirting	50/50% PE/ Cotton	3.91	50	95	100	20
8 T-Shirt Jersey	65/35% PE/ Cotton	4.82	35	85	100	20
15 Durable Press slack	65/35% PE/ Rayon	6.62	5	85	95	20
6 Untreated Slack	65/35% PE/ Cotton	6.99	40	75	90	20
1 Durable Press Slack	65/35% PE/ Cotton	7.07	30	85	100	20
C. MELT - SHRINK						
14 Taffeta	100% Nylon	1.67	0	0	5	13
12 Tricot	100% Nylon	2.47	< 5	15	25	20
2 Textured Woven Blouse	100% Polyester	2.19	0	0	0	8
3 Double Knit	100% Polyester	5.83	0	0	0	6
13 Tricot	100% Acetate	2.44	0	0	0	1
11 Tricot	80/20% Acetate/ Nylon	2.67	0	10	45	20
7 Jersey Tube Knit	100% Acrylic	2.64	10	10	35	20
D. WOOL						
20 Flannel	100% Wool	6.48	0	10	30	20

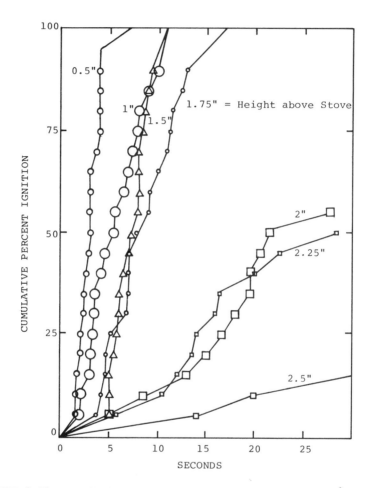

FIGURE 3-28   EFFECT OF HEIGHT ABOVE STOVE FOR 3.80 OZ/YD^2
COTTON FABRIC [R27]

171

ignition or melting times of the fabrics.

An approximate solution for this model (given in Reference [R-17] and in Section 3.2.1) yields:

$$\tau_{i,m} \cdot h_f \, \Delta T_f \simeq \rho \delta c (T_{p,m} - T_o) \qquad (3\text{-}27)$$

where $\Delta T_f = T_f - T_{avg}$ = Excess of flame temperature over fabric average temperature, T_{avg}, during the heating interval.

$T_{p,m}$ = Pilot ignition or melting temperature

h_f = Coefficient of convective heat transfer from flame to fabric, $W/cm^2 \cdot \,^\circ C$

T_f = Flame temperature[+] = 1850°C for a methane/air flame[R-17].

By assuming $T_{avg} = \dfrac{T_{p,m}}{2}$ and rearranging terms, eq. (3-27) becomes:

$$\tau_{i,m} \cdot h_f T_f \simeq \frac{\rho \delta c (T_{p,m} - T_o)}{\left(1 - \dfrac{T_{p,m}}{2T_f}\right)} \equiv E \qquad (3\text{-}28)$$

where E is a parameter that depends on the fabric thermophysical properties. (Note that E depends also on T_f, but very weakly, because $T_f \gg T_{p,m}/2$.)

The parameter E is tentatively proposed by the Overview Team to rank the GIRCFF fabrics with respect to the lower bound for the ignition (or melting) times under convective heating. This is done in Tables 3-7 and 3-8. Note that in a two-fold range for E, there is no consistent trend due to the fiber content.

[+]Flame temperatures as measured by GIT and corrected for radiation did not exceed 1400°C, as per Private Communication, August 12, 1974.

TABLE 3-7

FABRIC RANKING WITH RESPECT TO LOWER BOUND ON IGNITION
TIMES UNDER CONVECTIVE HEATING, AS PROPOSED BY OVERVIEW TEAM

Ranking Order	$E \equiv \rho \delta c (T_p - T_o)/(1 - \dfrac{T_p}{2T_f})$ $W \cdot s/cm^2$	GIRCFF Designation Number	Fiber Content
1	13	{ 6 { 15	65/35 PEC 65/35 PE/Rayon
3	12	{ 1 { 4 { 9	63/35 PEC Cotton Cotton
6	10	{ 8 { 19	65/35 PEC Cotton, FR
8	8.5	20	Wool
9	7.9	5	Cotton
10	7.4	16	50/50 PEC
11	6.5	18	Cotton
12	6.3	17	65/35 PEC
13	5.8	10	Cotton

TABLE 3-8

FABRIC RANKING WITH RESPECT TO LOWER BOUND ON MELTING
TIMES UNDER CONVECTIVE HEATING, AS PROPOSED BY OVERVIEW TEAM

Ranking Order	$E \equiv \rho \delta c (T_m - T_o)/(1 - \dfrac{T_m}{2T_f})$, $W \cdot s/cm^2$	GIRCFF Designation Number	Fiber Content
1	6.8	3	PE
2	6.3	7	Acrylic
3	5.1	12	Nylon
4	3.7	11	78/22 Acetate/Nylon
5	3.6	13	Acetate
6	2.7	2	PE
7	2.4	14	Nylon

Figure 3-29 shows that, indeed, for a constant convective heating flux, the parameter E correlates with the minimum ignition time measured by FMRC[R-17]. (This curve is essentially similar to Fig. 3-31 in Reference [R-17].) The correlation is not very good, as would be expected from the fact that the pyrolysis and melting characteristics of the fabrics have been neglected in determining the parameter E. Considerable deviation is noted for fabric #9, as was reported in Reference [R-17] and this unusual behavior remains inexplicable.

Because the convective heat-transfer coefficient, h_f, and hence the heat flux (from eq. 3-28), are unknown in the FMRC experiments, it is not possible to compare the magnitude of the measured ignition times to that calculated on the basis of eq. (3-28). This comparison would be a check on whether or not E represents a lower bound for ignition (or melting) time.

In light of the above discussion, the fabric parameter

$$E \equiv \frac{(\rho \delta c)(T_{p,m} - T_o)}{\left(1 - \dfrac{T_{p,m}}{2T_f}\right)}$$

(derived from a convective heat-transfer analysis of an inert thermally thin slab model), is tentatively proposed as a basis for ranking fabrics with respect to a lower bound on the ignition (or melting) times, under exposure to flaming ignition sources.

3.6   COMPARISON OF RADIATIVE AND FLAMING IGNITION

In this section, the results presented in previous sections for radiative and flaming ignition are compared. Then, a single fabric parameter is proposed to rank the GIRCFF fabrics regardless of the heating mode.

Figures 3-30 to 3-32 compare the experimental results in normalized form for the two heating modes for eight cotton, PE/C blend and synthetic fabrics. For the radiative ignition results,

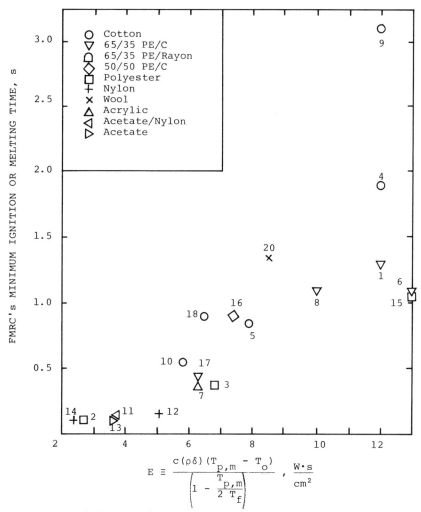

FIGURE 3-29   FMRC's MINIMUM RESPONSE TIME VS. FABRIC RANKING
PARAMETER, E

the power-law[†] dependence of the normalized time $(N_{Fo})_{i,m}$ on
the normalized heating intensity (q^*_{rad}) is well understood as
has been discussed in Section 3.4.

However, for flaming ignition of cotton and polyester-
cotton blends, the apparent downward concavity of the $(N_{Fo})_i$
vs. q^*_C relation remains unclear. (As illustrated in Fig. 3-32,
such downward concavity is not obtained in the case of two
synthetic fabrics for $1.5 \lesssim q^*_C < 6$.)

For cotton fabric #19, PE/C #8, and synthetics #2 and #12,
Figs. 3-30 to 3-32 show that the radiative and convective results
roughly merge, indicating that at any normalized heating intensity
the normalized ignition time is about the same regardless of the
heating mode. This is expected because it merely says that a
fabric will ignite (or melt) when it absorbs enough heat, by
radiation or convection, to reach its ignition[††] (or melting)
temperature. The deviation of cotton fabrics #5, #10 and #18
and PE/C #17 from this expected behavior remains unclear.

It should be noted, however, that at any heat flux (in
watt/cm^2) impinging on the fabric, the ignition (or melting)
time (in sec) will depend on the heating mode. As discussed
earlier, the heating mode determines the fraction (of the
impinging flux) absorbed by the fabric. Thus, while _normalized_
relations are very useful in representing large quantities of
data on the GIRCFF fabrics, only _non-normalized_ relations
indicate the hazard associated with each fabric under each
condition. In fact, the latter relations were used to determine
the fabric ranking parameters, A and E.

In Section 3.4.4 and 3.5.3, such parameters have been

[†] as given by eq. (3-22): $(N_{Fo})_{i,m} \simeq c(q^*_{rad})^n$

[††] in the case of flaming ignition, active radicals are brought
close to the fabric heated surface. This may lead to an
ignition temperature lower than that for radiative ignition
(as shown later in Table 3-9).

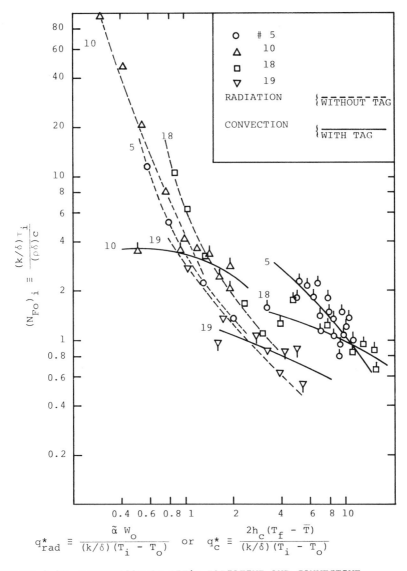

FIGURE 3-30   COMPARISON OF GIT's RADIATIVE AND CONVECTIVE
RESULTS FOR COTTON FABRICS

177

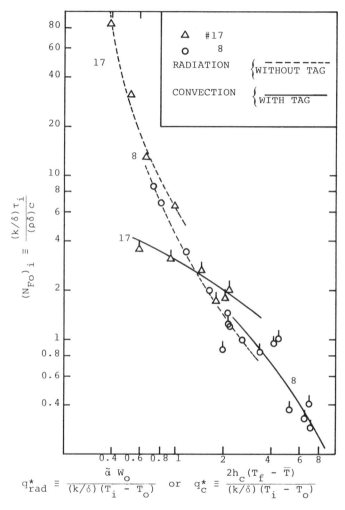

FIGURE 3-31 COMPARISON OF GIT's RADIATIVE AND CONVECTIVE
RESULTS FOR 65/35 POLYESTER/COTTON FABRICS

178

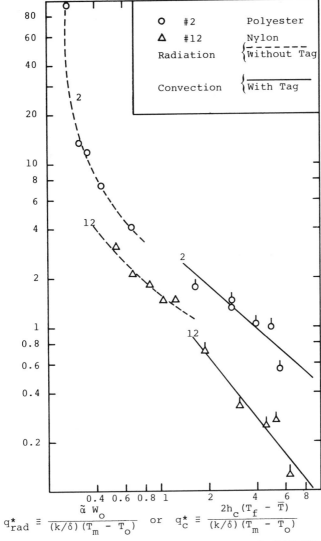

FIGURE 3-32 COMPARISON OF GIT's RADIATIVE AND CONVECTIVE MELTING
RESULTS FOR SYNTHETIC FABRICS

$$q^*_{rad} \equiv \frac{\tilde{\alpha}\ W_o}{(k/\delta)(T_m - T_o)} \quad \text{or} \quad q^*_C \equiv \frac{2h_c(T_f - \overline{T})}{(k/\delta)(T_m - T_o)}$$

179

tentatively proposed to rank the GIRCFF fabrics with respect to a lower bound on ignition (or melting) time under radiative and convective heating. The results for the two heating modes are hereby compared and a single parameter is proposed.

The lower bounds on ignition or melting times were found to be:

Under radiative heating, $A \equiv \dfrac{\rho \delta c (T_{i,m} - T_o)}{\tilde{\alpha}}$ (3-23)

Under convective heating, $E \equiv \dfrac{\rho \delta c (T_{p,m} - T_o)}{\left(1 - \dfrac{T_{p,m}}{2T_f}\right)}$ (3-28)

There is obviously an analogy between the results for the two heating modes. In general terms, the results can be expressed as:

(Ignition or Melting Time) x (Heat Flux) = (Fabric Parameter)

For each case, the fabric parameter is essentially a function of the fabric thermophysical properties. It also depends (somewhat) on the heating source through $\tilde{\alpha}$ or T_f, as explained in Sections 3.4.4 and 3.5.3. The numerator in each parameter represents the energy needed by the fabric to reach ignition or melting temperature. The denominator is a non-dimensional term which determines the fraction of energy absorbed by the fabric when subjected to a radiative or convective heat flux.

Figure 3-33 shows some correlation between the parameters A and E. (This is expected because the numerator is about the same in each term.)[†] Thus, a fabric that ranks high (or low)

[†]The only difference in the numerators is T_i in parameter A and T_p in E. This is obvious, because in the cases of radiative and flaming heat-sources, energy is needed to heat the fabric to the self-ignition temperature (T_i) and the auto-ignition temperatures (T_p), respectively.

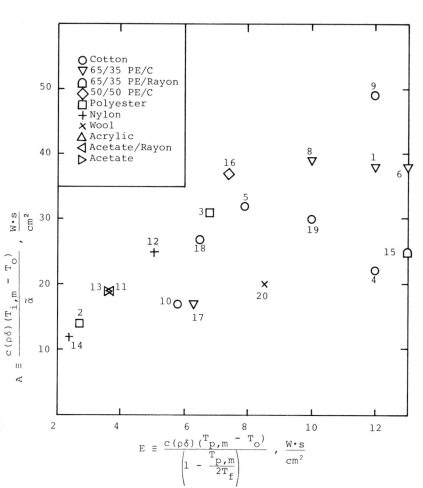

FIGURE 3-33   CORRELATION OF FABRIC RANKING PARAMETERS CHARACTERISTIC OF
RADIATIVE (A) AND FLAMING (E) IGNITION

under radiative heating will rank roughly in the same manner under convective heating. Furthermore, it is very important to note that, for each fabric, the value for E is smaller than that for A. This is also illustrated in Table 3-9. Thus, under the same "heating intensity", the lower bound on the ignition (or melting) time of a fabric is smaller for convective heating than for radiative heating. This is due to the following two reasons, in decreasing order of importance:

1. As illustrated in Table 3-9, the fabric absorptivity, $\tilde{\alpha}$, is much less than the non-dimensional excess flame temperature over fabric temperature (as averaged over the ignition interval). Thus, for any heat flux incident on the fabric, the energy fraction absorbed by the fabric is smaller for a radiative heat-transfer mode than for a convective mode.

2. For most fabrics the pilot ignition temperature, (T_p), is lower than the auto-ignition temperature, (T_i), as can be seen in Table 3-9. Thus, by definition, the energy needed by a fabric to reach pilot-ignition is less than that for auto-ignition. This distinction does not exist for melting fabrics, however.

Given that the nature of a heat source in an actual ignition accident is uncertain (as whether it is radiative or flaming), fabrics should be ranked with respect to their lowest bound on ignition (or melting) times, i.e., the parameter E. These rankings have been presented in Tables 3-7 and 3-8, Section 3.5.3.

3.7 SUMMARY OF RESULTS

The main results of this ignition study can be summarized as follows:

1. Ignition occurs well before the wearer could feel any pain due to thermal exposure to radiative or convective heat sources.

182

TABLE 3-9

COMPARISON RADIATIVE AND CONVECTIVE IGNITION PARAMETER

Fabric Number	$T_{i,m}$	T_p	$\tilde{\alpha}$	$\left(1 - \dfrac{T_{p,m}}{2T_f}\right)$	A $\dfrac{\rho\delta c(T_{i,m}-T_o)}{\tilde{\alpha}}$	E $\dfrac{\rho\delta c(T_{p,m}-T_o)}{\left(1 - \dfrac{T_{p,m}}{2T_f}\right)}$
	°C	°C			$\text{w.s}/\text{cm}^2$	$\text{w.s}/\text{cm}^2$
1	410	330	.35	.91	38	12
2M*	260		.18	.93	14	2.7
3M	250		.20	.93	31	6.8
4	290	290	.51	.92	22	12
5	310	310	.23	.92	32	7.9
6	410	340	.37	.91	38	13
7M	240		.72	.94	8.3	6.3
8	450	310	.35	.92	39	10
9	300	290	.24	.92	49	12
10	440	350	.39	.91	17	5.8
11M	230		.18	.94	19	3.7
12M	250		.19	.93	25	5.1
13M	230		.18	.94	19	3.6
14M	240		.19	.94	12	2.4
15	420	330	.60	.91	25	13
16	470	360	.24	.90	37	7.4
17	480	380	.43	.89	17	6.3
18	310	290	.24	.92	27	6.5
19	490	320	.50	.92	30	10
20	480	320	.61	.92	20	8.5

*M denotes Melting Fabrics

2. Ignition (or melting) times were carefully measured for the GIRCFF fabrics under a range of heat fluxes from radiative and convective heat sources and experimental results were compared with theoretical predictions. For the case of radiative ignition, there is qualitative agreement between test results and theoretical predictions for a thermally thin slab. The predicted times are always much lower, however, because the model neglects processes (such as moisture desorption, pyrolysis and convective losses) which would tend to retard ignition or melting. For the case of flaming ignition, the observed dependence of the normalized ignition time on the normalized heating intensity is not quite understood.

3. The ignition or melting time had a clearly defined minimum with respect to fabric location in the flame, occurring close to the luminous tip of diffusion flames or tip of the inner cone of pre-mixed flames.

4. The dependence of the minimum ignition times on non-fabric variables was investigated. Within the range used, burner-tube diameters and flame heigths (for both diffusion and pre-mixed flames) had no effect on the minimum ignition time; relative humidity and the proximity of a skin-simulant had a slight effect; and the air-methane mixture ratio had an important effect.

5. Within the range tested, the effect of porosity on the minimum ignition time appeared to be insignificant.

6. The real-life hazard of garment sleeves ignited by gas stoves has been simulated and the results obtained are comparable with strip tests.

7. A semi-empirical correlation of the data obtained in this study yields (for a specified radiative heat flux) an expression for the ignition (or melting) times in terms of a fabric parameter,

$$
D \left(\equiv \frac{\rho \delta c (T_{i,m} - T_o)^{3/2}}{\tilde{\alpha}^{3/2}} \; (k/\delta)^{1/2} \right) ,
$$

based on the fabric thermal and physical properties. This parameter ranks these properties with respect to their importance to ignition (or melting) in the following order:

    (a) the ignition or melting temperature and the absorptivity,

    (b) the weight and the specific heat,

    (c) the thermal conductance.

8. For radiative heating, a <u>lower bound</u> for the measured ignition (or melting) times can be given by the fabric-parameter,

$$
A \left(\equiv \frac{\rho \delta c (T_{i,m} - T_o)}{\tilde{\alpha}} \right) ,
$$

resulting from a heat-transfer analysis of an <u>inert thermally-thin slab</u> model.

9. For convective heating, a <u>lower bound</u> for the measured ignition (or melting) times can be given by the fabric-parameter

$$
E \left(\equiv \frac{\rho \delta c (T_{p,m} - T_o)}{1 - \left(\frac{T_{p,m}}{2T_f}\right)} \right) ,
$$

also resulting from heat-transfer analysis of an <u>inert</u> <u>thermally-thin slab</u> model.

10. The parameter D can be useful in systematically modifying the ignition (or melting) behavior of fabrics.

11. The parameters A and E are tentatively proposed by the Overview Team to rank the fabrics with respect to a lower bound on the "ease of ignition (or melting)" under radiative and convective heating respectively.

12. Given that the parameter E is always smaller than A, it can be considered as the lower bound on the "ease of ignition (or melting)" under radiative <u>and</u> convective heating. Thus, to rank the fabrics regardless of the heating mode, the parameter E should be used.

CHAPTER IV

FLAME SPREADING

4.1   INTRODUCTION   188

4.2   FLAME PROPAGATION IN FABRIC STRIPS   195
      4.2.1   General Observations on the Effect of Fiber
              Composition   196
      4.2.2   Samples Placed in a Horizontal Plane   199
      4.2.3   Samples Placed in an Inclined (45°) Plane   201
      4.2.4   Samples Placed in a Vertical Plane   203
      4.2.5   Comparison of Orientation Effects   209
      4.2.6   Flame Spreading in Fabric Composites   209

4.3   FLAME SPREADING IN GARMENTS   213
      4.3.1   Single Garments   213
      4.3.2   Garment Combinations and Design   220
      4.3.3   Comparison of Single Garments and
              Combinations   222
      4.3.4   Observations on the "Ease of Extinguish-
              ment"   226

4.4   COMPARISON OF FLAME SPREAD RESULTS ON FABRIC STRIPS
      AND ON GARMENTS   226

4.5   EFFECTS OF FABRIC PARAMETERS (NOT CONSIDERED BY
      GIRCFF) ON FLAME SPREADING   227

4.6   OVERVIEW   232

CHAPTER IV

FLAME SPREADING

4.1 INTRODUCTION

As was pointed out in Chapter I, "Hazards of Flammable
Fabrics", the work plan of the GIRCFF has focused on two probabilistic elements in the series of events which lead to human
injury from burned clothing: (a) the probability of ignition of
a given fabric or garment, given a specific thermal exposure,
and (b) the probability of burn injury to the wearer of the garment following fabric ignition. It was pointed out that the
concept of burn probability actually includes the probability
of flame spread taking into consideration the fabrics as combined
into garments, the garment design and fit, and also wearer reaction.
When the fabric in a garment is ignited, the flame may continue
to propagate, or it may move a short distance and then self-
extinguish. Furthermore, if the flame spreads, the probability
of thermal injury or burns may be considered in terms of the
heat transferred from the flaming garment to the skin.

It seems logical, therefore, to subdivide the phenomena
leading to burn injury into a series of events beginning with
ignition of the fabric and ending with heat transfer from the
burning fabric to the skin. As has been pointed out, injury from
flammable fabrics is a result of transfer of heat to the skin in
excess of human tolerance [34]. Most commonly used commercial
fabrics have such large heats of burning that the heat transfer
(under the conditions prevailing in fabric fire accidents) is
generally in excess of this tolerance limit. This has been shown
to be the case for the 20 fabrics tested in this GIRCFF program.
Thus, there appears to be some justification for the comment
heard from the medical profession that a square centimeter of
burned fabric corresponds to a square centimeter of burned skin.
However, as Chouinard points out [34], "The intricate nature of
energy generation and transfer from fabrics of different types
can cause great differences in both the amount of energy transferred and the total body area involved".

188

If every square centimeter of fabric burned in the proximity of the body generates enough heat of combustion and transfers enough of that heat to cause damage to the body, then the probability of injury would appear to be determined primarily by the probability of flame propagation in various fabrics and in various combinations of garments. This view is one which appears to dominate in the development of test methods for flammable fabrics, for most tests focus on ignition and propagation of flame rather than upon the level of heat transfer from a burning fabric or garment.

Nonetheless, the GIRCFF program was also concerned with the heat transfer from fabrics which had been ignited and through which a flame was propagating. Considerable effort was devoted to the measurement of heat flux and to its correlation with epidermal necrosis, as well as to the development of skin simulants and heat flux indicators which would permit an evaluation of skin damage without the necessity for animal specimens. As it turned out, most experiments conducted in the two laboratories concerned with this phase of the program, namely, the Fuels Research Laboratory of the Chemical Engineering Department, M.I.T., and the Gillette Research Institute, recorded the position of the flame as a function of time and thus inherently determined rate of flame propagation across the fabric and garment combinations used in the experiments.

In view of the fact that most flammability tests focus on ignition and flame propagation, and in light of reports in the literature on the effect of fiber content and fabric structure on propagation rates, it is considered desirable to discuss the results obtained on flame propagation in the GIRCFF program before proceeding with the discussion of heat transfer and burn injury data.

According to Bernskiold [12], when a textile fabric is exposed to an igniting source, a variety of phenomena may follow: (a) the fabric does not ignite and gases formed by pyrolysis of the fibers do not burn in the igniting flame, (b) the fabric does not ignite, but the gases formed do burn in the igniting flame, (c) the fabric ignites but burns only for a

189

short interval after removal of the igniting flame, (d) the
fabric ignites and continues to burn after removal of the
igniting source. Bernskiold [12] suggests the use of a double
coordinate system to record the behavior of a textile product
during the initial ignition period and during the subsequent
period of burning; length of the destroyed specimen is plotted
as a function of the igniting time and of the burning time. In
such a diagram, the zero value of the burning time actually
corresponds to the ignition time.

The exact boundaries of the burning area, the flame area and
the charred area should be determined in any attempt to characterize propagation of the flame. An example of this [22] for a
vertically suspended specimen is indicated in Fig. 4-1. And it
is suggested that ignition and flame propagation can be described
in terms of changes in the following quantities: distance between
the outer boundary of the burning zone in the fabric and the
ignition point (y_b), and distance between the flame front possibly
extending beyond the burning zone and the ignition point (y_f).
To provide for measurement of sideways movement of the flame, one
can designate the distance (y_{bs}) between the ignition point and the
side-boundary of the burning zone. An example of a typical
propagation record is shown in Fig. 4-2. This diagram may be used
for the case where the products of pyrolysis induced by the
igniting flame are not ignited, and the burning zone (y_b) will
tend to reach a constant value during the igniting flame time.
When combustible gases formed in pyrolysis burn during the ignition
period, the length of the destroyed area of the specimen will be
larger than for the case when the gases do not burn.

Graphs illustrating the length of the destroyed fabric area
for these two cases are shown in Fig. 4-3. In case (A1) where
the combustible gases resulting from pyrolysis of the fabric do
not ignite, it may be expected that burning will not proceed when
the ignition source is removed. For case (A2), however, where
the pyrolysis products do burn, the fabric may either extinguish
upon removal of the igniting flame, or continue to burn, as is
shown in Fig. 4-4 as cases (B1) and (B2), respectively.

Thus, there are several parameters which can be measured

190

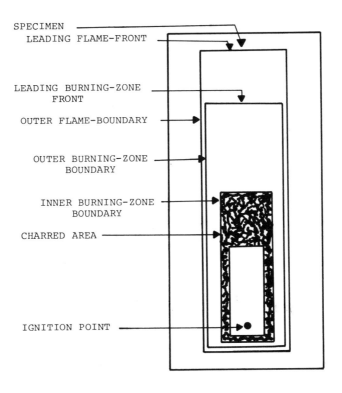

FIGURE 4-1   DESCRIPTION OF A BURNING VERTICALLY-SUSPENDED
SPECIMEN   [12]

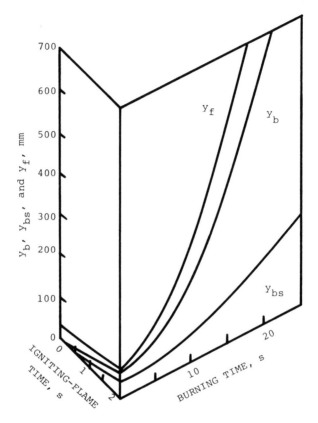

FIGURE 4-2   RELATION BETWEEN FRONT DISTANCES
AND TIME AS SHOWN BY IGNITION-TIME AND
BURNING-TIME DIAGRAMS   [12]

y_f = DISTANCE BETWEEN LEADING FLAME-FRONT AND IGNITION POINT

y_b = DISTANCE BETWEEN BURNING-ZONE BOUNDARY AND IGNITION POINT

$y_{b,a}$ = DISTANCE BETWEEN BURNING-ZONE SIDE-BOUNDARY AND IGNITION POINT

192

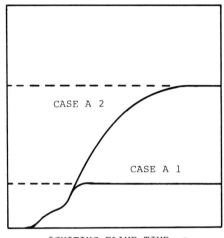

IGNITING FLAME TIME, s

CASE A 1 - PYROLYSIS GASES DO NOT BURN IN
THE IGNITING FLAME

CASE A 2 - PYROLYSIS GASES BURN IN THE
IGNITING FLAME

FIGURE 4-3   LENGTH OF THE DESTROYED SPECIMEN AS A FUNCTION
OF THE IGNITING FLAME TIME   [12]

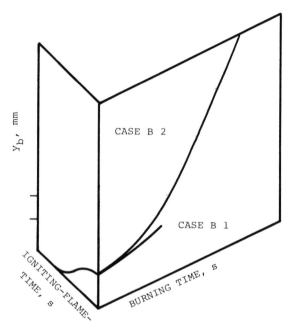

y_b, mm

CASE B 2

CASE B 1

IGNITING-FLAME-
TIME, s

BURNING TIME, s

CASE B 1 - BURNS AWHILE
CASE B 2 - BURNS

FIGURE 4-4   LENGTH OF THE DESTROYED SPECIMEN AS A FUNCTION
OF THE IGNITING-FLAME TIME AND BURNING TIME   [12]

in the testing of flammability of textile fabrics. One can
bracket the time period within which ignition of a given fabric
subjected to a given exposure condition will take place by
changing the igniting exposure time over a wide range. This
can be termed the ignition time. One may determine the constant
rate of flame propagation, assuming that the fabric geometry and
orientation is such as to permit the development of a steady
rate of flame propagation. (It is conceivable that in some cases
steady flame propagation may not be attained.) One may determine
the heat transfer from the flame to an adjacent surface, such as
the skin, as the flame front and the burning region of the fabric
pass a given skin position. Finally, one may determine the
extinguishment time--from the occurrence of ignition to self-
extinguishment (if the latter occurs). Ignition time measurements and predictions obtained in the work of GIT and FMRC
have been discussed in Chapter III. This chapter will focus on
data obtained for the constant or average velocity of propagation
for the cases of steady or unsteady flame spreading respectively.
The averaging is usually done over the length of the specimen.
Heat transfer from the burning fabrics to adjacent surfaces will
be discussed in Chapter V.

4.2 FLAME PROPAGATION IN FABRIC STRIPS

The study of flame propagation in selected GIRCFF fabrics
was undertaken by the Fuels Research Laboratories at M.I.T.
Burnings of fabrics were conducted with a provision of uniform
tensile loading of a fabric strip held over a surface used as a
skin simulant, of controlled spacing between the fabric and the
simulant, and of controlled orientation of the fabric-skin
combination with regard to the horizontal. The fabric specimen
was cut to a size 20" x 12" and clamped along the longer edge.
The other longer edge of the specimen was slipped between a
second set of long clamps and the loose edge was then dead
loaded. The second set of clamps was then tightened so that
the fabric specimen was subjected to a lateral load of about

1 lb. force/foot length. A Transite simulant was placed under
the fabric and parallel to it, leaving a 20" x 6" fabric specimen stretched uniformly over a 24" x 7" x 3/8" Transite
simulant. Asbestos tape was used to cover the innerside of
the clamp bars so as to minimize quenching effects along the
specimen edges. Finally, the spacing between the fabric and
the simulant was adjusted to any desired value. And the entire
assembly was tilted at the desired angle to the horizontal.

A multi-hole gas burner served as the ignition source and
consisted of a hollow brass tube 6" in length, 1/2" in diameter
with fine holes drilled in it. This tube was mounted on the
top of the vertical tube of a Bunsen Burner and provided a
uniform line source of ignition. Before testing, the fabric
and skin simulant were allowed to reach equilibrium in a controlled environment chamber. The fabric was then ignited at
one end and temperature measurements were made along the skin
simulant as the flame propagated over the fabric. The movement
of the flame was observed visually with time as a basis for
calculating average speeds of flame propagation.

4.2.1 General Observations on the Effect of Fiber Com-
       position

The qualitative features observed for flame spreading
varied significantly with fiber composition. The most important observations are summarized below for the fibers included
in the GIRCFF fabric series:

1. The cotton fabrics (untreated) all sustained ignition
    and, in most cases, propagated a uniform flame. "The
    heavier fabrics produced whitish smoke during burning, but
    most of the lighter fabrics produced relatively little
    visible smoke. In horizontal burning, flame propagation
    was generally sufficiently slow that flame propagation
    velocity could be measured fairly accurately by visual
    observation. In horizontal burning of heavier fabrics,
    however, flame propagation was irregular and became
    increasingly so as the fabric-skin spacing was reduced
    from 1/2" to 1/4" when the flame died out because of in-
    sufficient air supply and quenching by the skin simulant.

196

In horizontal burning, the height of the flame reached from 1/2" to 1" approximately. The burnt fabric usually fell in the form of a hot char on the underlying simulant surface, and this char often continued to glow for some time.

"In 45° upward and 90° upward burning, flame propagation was quite rapid, and only a crude estimation of average flame speeds could be made by measuring the time to burn the total length of the fabric specimen. Flame height was about 1' to 1½'. The fire retardant-treated cotton fabric (#19) did not propagate a flame in any test."

2. The polyester/cotton blends "in general, produced large amounts of black smoke above the flame and white fumes in the fabric-skin air gap, particularly in horizontal burning. In all cases, the fabric melted slightly ahead of the flame with a sizzling liquid appearance. Melt dripping was pronounced, particularly in horizontal burning, when drops of melt would frequently fall on the skin simulant surface and either continue to burn there, or solidify into hard deposits which had to be scraped off the surface with a blade. Furthermore, the heated fabric ahead of the flame would soften and sag, producing irregular burning.

"Similar phenomena were also observed in 45° upward and 90° upward burning except that the frequency of hot melt drops contacting the simulant surface was less in the former orientation and almost absent in the latter. The polyester-rayon sample (#15) behaved essentially like the polyester/cotton blends in all orientations".

3. "The 100% synthetics (with the exception of the acrylic fabric), did not propagate a flame or, if they did, it was so irregular that a statistical study would be required to produce meaningful results. The polyester samples #2 and #3 ignited when the flame was brought into contact with the fabrics; the heavier double knit sample (#3) propagated a very irregular, narrow flame, leaving

197

most of the specimen unburnt. The lighter sample (#2),
however, melted at the ignition point and the melting region separated from the remainder of the sample, leaving
it unburnt.

"The light-weight nylon samples (#12 and #14) melted
rapidly and shrank away from the ignition source, i.e.
burning was not sustained in the absence of an external
igniting flame.

"Acetate fabric (#13) burned irregularly upon ignition
but did not propagate a flame in horizontal orientation.
In 45° upward burning, the flame propagated in a zig-zag
pattern.

"The acetate-nylon blend (#11) burned irregularly,
producing smoke. The fabric melted ahead of the flame
which propagated in numerous directions, leaving islands
of unburned fabric. In the horizontal orientation, the
unburnt pieces, melt, and burnt char flakes fell on the
simulant surface; the melt was soft and did not stick to
the surface".

The acrylic (#7) burned with a characteristic odor.
It produced black smoke and gave a high flame (about 3")
in a horizontal orientation. The fabric melted ahead of
the flame, the sticky melt dropped on the simulant below
and sometimes continued to burn there. In 45° upward
orientation, however, the fabric broke into small pieces
while burning; these pieces stuck to the simulant and
sometimes continued burning.

4. "The 100% wool fabric (#20) ignited and gave an
appearance of melting, but the flame extinguished itself
in the horizontal orientation. In the 45° upward
orientation, the fabric burned uniformly, but the flame
constricted to a narrow width approximately in the middle
of the specimen and the burnt cloth tended to curl upward.

On the basis of these observations, the M.I.T. Fuels
Laboratory group decided to confine their experiments to
the all-cotton fabrics and the polyester/cotton blends,

since it appeared that the remaining fabrics, if burned
alone, could not be expected to give reproducible results.

4.2.2 Samples Placed in a Horizontal Plane

1. Effect of Spacing

It has been suggested (ref. Section 4.2.1) that the
spacing between the burning fabric and the skin simulant
would have an important effect on the rate of flame propagation. As the M.I.T. group pointed out, spacing of
garments from the body may vary from 0 to more than 1",
depending on the type of garment worn and the part of
the body which it covers. Accordingly, the M.I.T. group
studied the effect of fabric/skin simulant spacing on
flame propagation, and on predicted burn damage for three
of the GIRCFF fabrics, specifically one cotton fabric and
two polyester/cotton blends. Five spacings were examined
experimentally under controlled environmental conditions
(70°F and 65% relative humidity). The average flame
speeds measured, under these conditions of burning, are
shown in Table 4-1. None of the fabrics tested propagated
a flame when the spacing was limited to 1/8", indicating
either an insufficiency of air for combustion or a quenching
effect of the adjacent skin simulant. At the 1/4" spacing,
the 3.7 oz. cotton flannel fabric (#18), with rather coarse
low-twist yarns and a napped surface, propagated a somewhat
irregular flame, while the polyester/cotton blends which
were made of finer, more highly twisted yarns did not
sustain combustion.

The data summarized in the Table indicate that, in
the range studied, average flame speed increases with
increased spacing, as might be expected due to increased
air supply or reduced heat loss to the adjacent Transite
surface. Furthermore, the Table shows that the maximum
thermal dosage to skin occurred at a 1/2" spacing. Consequently, this spacing was chosen for the experiments
designed to investigate the effect of parameters related
to flame spreading.

Table 4-1

Effect of Fabric-Skin Spacing, Horizontal Burning

(65% Relative Humidity at 70°F) [R25]

Fabric Number	Fabric Description	Spacing (inch)	Average Flame Speed (cm/min)	Thermal Dosage to Skin (cal/cm^2)
18	100% Cotton 3.69 oz/yd^2	1/8	no propg'n	---
		1/4	5.5	5.5
		3/8	8.6	4.1
		1/2	10.4	6.3
		3/4	20.0	3.6
17	PE/Cotton 65/35 2.53 oz/yd^2	1/8	no propg'n	---
		1/4	no propg'n	---
		3/8	16.7	2.7
		1/2	20.5	3.4
		3/4	42.9	1.7
6	PE/Cotton 65/35 6.99 oz/yd^2	1/8	no propg'n	---
		1/4	no propg'n	---
		3/8	7.5	3.9
		1/2	8.3	6.5
		3/4	12.3	3.3

2. Effect of Fabric Weight/Unit Area

The measurements of average flame spreading speeds for cotton and polyester/cotton fabrics at a fabric/simulant spacing of 1/2" are shown in Fig. 4-5. Here it is seen that there is a linear relationship with a negative slope between average flame propagation and weight-per-unit area of the fabric (given in ounces/square yard). The data for the polyester/cotton lie above those for 100% cotton fabrics and it would appear that, for a given weight fabric, polyester/cotton propagates the flame at a rate of about 2 centimeters/minute faster than 100% cotton fabrics.

4.2.3 Samples Placed in an Inclined (45°) Plane

When the fabric and skin simulant assembly was inclined at 45° and ignition started at the bottom of the fabric so that the flame propagated upward, the orientation corresponded to that used in some standard methods of testing fabrics for wearing apparel and possibly also to the upward flame propagation noted frequently in accidents involving apparel. One-half inch spacing between fabric and skin simulant was used in all the 45° upward burning tests and the environment was set at 70°F and 65% RH. Since the flame spread is quite rapid under these conditions, estimates of average flame speeds were made by measuring the time required for burning of the entire 20" length of the specimen. As in the horizontal burning tests, measurements were complicated in some instances with hot char and melt drips falling on the simulant. Also, fabrics were observed to soften and sag, thus altering the spacing between the fabric and the simulant in the course of the experiment, and causing slower or "bad burning". This characterization of "bad burning" as against "good burning", when the original geometry is maintained, has been used from time to time by the M.I.T. group in describing results. It was reported, for example, that some fabrics, when ignited uniformly across the full width along their bottom edge, tended (because of sagging) to burn faster on the sides than in the center. This caused even more sagging of the middle portion of the fabric towards the skin simulant,

201

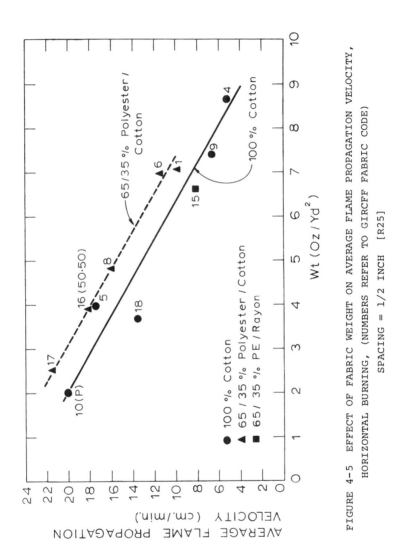

FIGURE 4-5 EFFECT OF FABRIC WEIGHT ON AVERAGE FLAME PROPAGATION VELOCITY,
HORIZONTAL BURNING, (NUMBERS REFER TO GIRCFF FABRIC CODE)
SPACING = 1/2 INCH [R25]

and the sample either burned even slower or was extinguished. The 100% cotton fabrics generally exhibited "good burning" in the 45° upward burning tests and excellent reproducibility was obtained in replicate runs.

The polyester/cotton blends, on the other hand, manifested irregular flame propagation and five-to-seven repetitive tests had to be carried out and averaged for each of these fabrics. More rapid burning of the specimen along its sides occurred due to the sagging and it was found necessary to ignite the middle of the fabric edge with a computer card strip or with a match. The data obtained in this manner for the polyester/cotton blends are shown in Fig. 4-6. Average flame speed is much greater than for horizontal burning, but in this orientation also, it is inversely related to fabric weight. In this case, it would appear that the cotton fabrics exhibit slightly greater average flame speed than the polyester/cotton blends. However, the difference is small and perhaps not significant in view of the experimental uncertainties.

4.2.4 Samples Placed in a Vertical Plane

The same set of fabrics was tested in 90° upward (vertical) burning, an orientation which is obtained frequently in serious burn accidents. Again, average flame speeds were measured by visual observation. Cotton burned well with considerable char formation particularly in the case of heavier fabrics. The polyester/cotton blends frequently softened and sagged approaching the simulant surface during burning. As for 45° upward burning, the sides of the sample burned faster than the middle in the case of blend fabrics, and this appeared to draw the sample closer to the simulant, sticking to it in some places and burning slowly. Again, the differentiation between "good" and "bad burning" is based upon this kind of specimen distortion. The data for the 100% cotton series of fabrics and for the polyester/cotton blend series are shown in Fig. 4-7. The inverse relationship between flame speed and fabric weight was again demonstrated, and the polyester-cotton fabrics appeared to burn slightly faster than the all-cotton

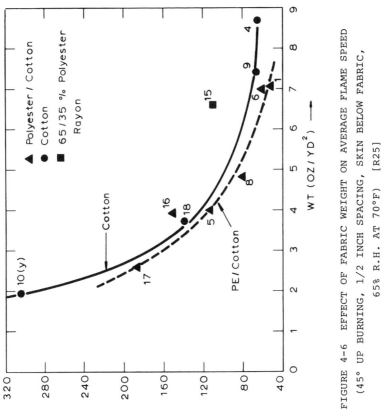

FIGURE 4-6 EFFECT OF FABRIC WEIGHT ON AVERAGE FLAME SPEED
(45° UP BURNING, 1/2 INCH SPACING, SKIN BELOW FABRIC,
65% R.H. AT 70°F) [R25]

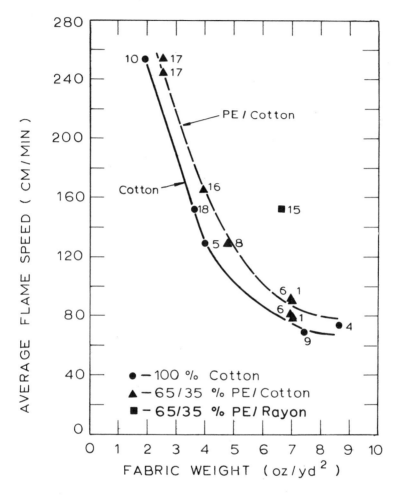

FIGURE 4-7   EFFECT OF FABRIC WEIGHT ON AVERAGE FLAME SPEED;
90° UP BURNING, 1/2 INCH SPACING
(NUMBERS REPRESENT GIRCFF FABRIC CODE)   [R25]

fabrics, as in the case of 45° upward burning.

Effect of Fabric-Skin Simulant Spacing in Vertical Burning

Since the 90° orientation may be assumed to correspond most closely to the situation of textile fabrics used in garments, it was considered desirable to study the effects of spacing in this configuration also. For this purpose, two cotton fabrics were used with four different spacings in 90° upward burning experiments. These fabrics propagated a uniform flame for spacings ranging from 1/8" to 1/2". (Note that in horizontal burning uniform propagation could only be achieved at spacings above 1/4" [Section 4.2.2].) The results of these tests are shown in Fig. 4-8 for the 2 oz. cotton fabric sample #10 and in Fig. 4-9 for the 3.7 oz. cotton flannel sample #18.

These figures also include data for thermal injury which will be discussed fully in Chapter V. At this time, it suffices to point out that thermal injury calculated for these two fabrics is highest at the minimum spacing, namely 1/8", where the average speed of flame propagation is lowest. Injury then decreases with increasing spacing and reaches a minimum at a spacing of 9/32". Subsequently, it increases as the space is widened and goes through another maximum at about 15/32".

As reported in Chapter V, the M.I.T. group explained this phenomenon as follows. "The factors affecting the total thermal dosage and injury to the skin are, (1) radiation from hot gases and fabric which is directly proportional to the fourth power of the absolute temperature and inversely pro-portional to the square of the spacing, (2) convection and conduction through the gas phase which is directly proportional to the temperature of the hot gases above the skin and is a direct function of gaseous recirculation, (3) condensation of steam and pyrolysis products, which becomes significant at a certain threshold spacing and increases with the decrease in spacing to a critical spacing below which flame spreading cannot be sustained, (4) exposure time, which is inversely proportional to flame speed and thus decreases with increase in spacing." The effect is a complicated one, and speed of flame propagation

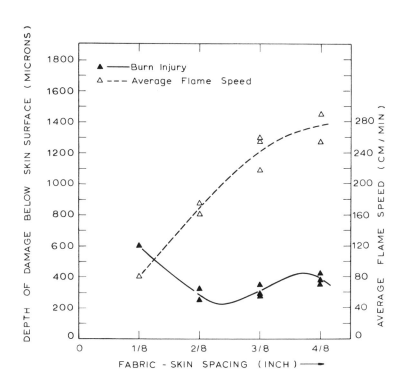

FIGURE 4.8   EFFECT OF SPACING ON BURN INJURY AND AVERAGE
FLAME SPEED, 90° - UP BURNING, #10 FABRIC
(2.0 oz/yd²)   [R25]

207

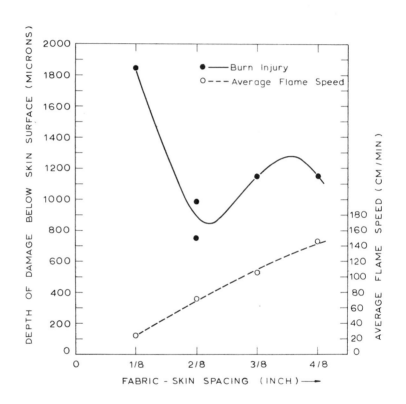

FIGURE 4-9   EFFECT OF SPACING ON BURN INJURY AND AVERAGE
FLAME SPEED, 90° - UP BURNING, #18 FABRIC
(3.7 oz/yd²) [R25]

is but one factor in the complex interactions that determine
the severity and the extent of injury from burning fabric in
a given system geometry.

4.2.5 Comparison of Orientation Effects

The effect of orientation of the fabric and skin simulant
on the speed of flame propagation is summarized in Table 4-2
for 100% cotton and for polyester/cotton fabrics in the GIRCFF
series. The data show that average flame speed decreases with
increasing fabric weight for both fiber compositions. Generally,
the flame velocity is somewhat higher for the 90° orientation
than for the 45° orientation in upward burning, but there are
some exceptions. On the other hand, the flame velocities in
both upward orientations are an order of magnitude higher
(that is, ten times as high) as observed in the horizontal
burning tests.

4.2.6 Flame Spreading in Fabric Composites

Since few fabrics are ever used in single layers, the Fuels
Research Laboratory group at M.I.T. considered it desirable to
examine the burning characteristics of fabrics as composites.
The group observed that meaningful, repeatable burning tests on
single layers of polyester fabric, nylon, wool, acetate and
acrylic could not be carried out. Accordingly, it was decided
to test these fabrics as composites with GIRCFF fabric #5, a
100% cotton knit T-shirt material, commonly employed in undergarments worn by men.

The fabrics were mounted together in the testing apparatus
essentially as described for the single layer fabric. Tests
were performed for 45° upward burning, and for 45° downward
burning, with the skin simulant positioned below the fabric
and aprallel to it. In this configuration, the burn tests
proceeded fairly well and it was noted that all of the synthetics
melted and shrank ahead of the flame front. Uniformity of
burning was another matter. The 100% wool, together with the
cotton undercloth, burned uniformly. However, the middle portion of the composites, including 100% synthetics, sagged, and
therefore burned more slowly than at the sides. The behavior
was similar to that described for the polyester/cotton fabrics
tested alone. For this case, a slight change in procedure was

209

TABLE 4-2

Effect of Fabric-Skin Orientation on Flame Speed

1/2 Inch Spacing [R25]

Fabric Number	Weight (oz/yd^2)	Average Flame Speed (cm/min)		
		Horizontal	45° Up	90° Up
Cotton				
10	2.0	20.0	304.8	254.0
18	3.7	13.5	139.0	152.4
5	4.0	17.5	113.0	130.0
9	7.4	6.7	66.3	69.3
4	8.7	5.3	67.2	73.5
PE/Cotton				
17	2.5	24.1	194.7	249.0
16	3.9	18.0	152.4	165.0
8	4.8	16.0	91.8	128.0
15	6.6	8.1	114.4	152.4
6	7.0	11.5	56.0	86.3
1	7.0	9.8	50.8	82.7

introduced and a 1/2" computer card was attached to the lower edge (or upper edge, depending on the test) in the middle portion of the specimen so as to initiate preferential ignition of the middle portion.

The results of 45° upward burning tests on these composites are summarized in Table 4-3. The last line in this Table represents data for fabric #5, the 100% cotton knit T-shirt fabric tested alone. All other data are for the indicated fabric burned on top of the cotton T-shirt (fabric #5) as a two-layer composite. The data indicate that all the composites burned more slowly than the all-cotton knit fabric, and the wool composite burned at less than one-half the speed of fabric #5 alone.

The burning of the 100% thermoplastic fabrics, nylon (#14 and #12), polyester (#2 and #3), acrylic (#7), acetate/nylon (#11), and acetate (#13) proceeded step-wise. The problem of melting and shrinking of the upper synthetic fabric layer caused the thermoplastic component to build up locally to the point where it could no longer shrink away from the flame and, at this point, the accumulated melt would ignite. Then the sequence of melting, shrinking and ignition would repeat. In some instances it was thought that molten polymer from the top layer fabric would impregnate the under cloth, reduce its porosity and significantly alter its combustion behavior.

To avoid this step-wise burning, another series of experiments were conducted in which the composites were burned with the flame propagating downwards at 45°. Under these conditions, it was found that all of the fabrics burned with a steady flame. The thermoplastic fabrics generally melted ahead of the flame front forming small globules which rested on the undercloth structure and appeared to feed the flame uniformly. This was a marked effect, particularly for the case of the polyester fabrics. With nylon as the top layer fabric, the composite structure did not exhibit a uniform flame front because the melt tended to flow towards the middle of the specimen forming few globules which fed the flame, and the center portion of the fabric burned faster than the sides.

211

TABLE 4-3

45° Up Burning of Wool and Pure Synthetics as Composites
with Fabric #5 as Undercloth (100% Cotton, T-Shirt);
1/2 Inch Spacing (65% Relative Humidity at 70°F) [R25]

Fabric Type	Top Fabric Number	Weight of Top Fabric (oz/yd^2)	Mode of Ignition	Type of Burning	Average Flame Speed (cm/min)
100% nylon	14	1.67	Burner	bad	87.0
	14	1.67	Burner	fair	78.2
	14	1.67	1" Computer Card	good	72.6
	14	1.67	1" Computer Card	good	65.0
100% nylon	12	2.47	1" Computer Card	fair	72.6
	12	2.47	1" Computer Card	good	58.6
	12	2.47	1" Computer Card	good	59.8
100% polyester	2	2.19	Burner	bad	95.3
	2	2.19	Burner	bad	87.0
	2	2.19	1" Computer Card	fair	70.0
	2	2.19	1" Computer Card	good	75.0
	2	2.19	1" Computer Card	good	75.0
100% polyester	3	5.83	Burner	bad	--
	3	5.83	2" Computer Card	good	44.8
	3	5.83	2" Computer Card	fair	57.5
100% acrylic	7	2.64	Burner	bad	67.7
	7	2.64	Burner	good	61.0
Acetate/nylon	11	2.67	1" Computer Card	good	69.3
	11	2.67	1" Computer Card	good	66.3
Acetate	13	2.44	Burner	good	78.2
100% wool	20	6.48	Burner	partly unburned at end	46.9
	20	6.48	Burner	good	49.2
	20	6.48	Burner	good	50.8
100% cotton	5 (control)	3.97	Burner	good	113.0

212

Thus, it was concluded by the Fuels Research Laboratory group at M.I.T. that the synthetics could be made to propagate the flame when combined with cotton fabric and that it would be possible to rate the fabrics with regard to flame spread and heat transfer in this manner. Downward movement of the flame at 45° orientation was found to cause more damage than the 45° upward burning flame. These observations will be discussed in Chapter V. For the study of flame propagation, however, upward burning was preferred as a testing mode because it was thought to be more realistic.

4.3 FLAME SPREADING IN GARMENTS

The Gillette Research Institute studied flame propagation characteristics of the GIRCFF fabrics in garment form, using mannequins instrumented with calibrated heat flux sensors. The results obtained were recorded in terms of heat input at the sensor locations, and also in terms of percentage area of the torso which would have suffered second-degree burns, or worse. Since these results were recorded for specified burning times, they provide a basis for judging relative speeds of flame propagation in garments (dresses) made from the GIRCFF fabrics. In addition, tests were run in similar manner with various combinations of garments, such as might be used in normal practice (e.g. dresses with slips). Some evaluation of the effects of garment design and fit was also undertaken through the use of a belt with some of the dresses, and in some tests carried out with slacks tailored in tight and loose configurations. The dresses were generally ignited at the front of the hem and, in the case of slacks, ignition was started at the right knee.

4.3.1 Single Garments

The results of the Gillette Research Institute investigations are summarized below:

1. Cotton and polyester/cotton dresses burned extensively causing injury over 70% or more of the torso and head when allowed to burn to completion.

2. Dresses made from nylon, polyester, NOMEX, or fire

213

resistant cotton fabrics did not sustain combustion when
tested as single layer garments even when ignition was
attempted by means of a large tape attached to the hem.
3. Dresses made from 100% cotton and from polyester/
cotton blend fabrics of comparable weight burned to the
same ultimate area but, at comparable fabric weight,
cotton was observed to burn more quickly.
4. The pattern of flame spread for cotton and polyester/
cotton blend dresses ignited at the front of the hem,
was described as follows. The flame spread up the front
of the dress from the point of ignition, but did not always
reach the front of the shoulders. Flames then spread
relatively slowly around to the back of the dress, often
cutting off large portions of unburned fabric which would
fall away, leaving some sensor areas unaffected. Figure
4-10 shows examples of the flame-spreading patterns on
dresses tested as single layers. The contours in Fig. 4-10
show the position of the flame front at given times.
Upward movement of the flame from the point of ignition
along the front of the dress is more rapid than the
sideway movement to the back of the dress. This rapid
movement of the flame front immediately after ignition
is also indicated in Fig. 4-11, which is a plot of the
area of torso burned (in percentage) vs. time after
ignition. This figure also indicates typical replicate
results obtained in tests carried out with single layer
dresses made from fabric #17, a 2.5 oz. polyester/cotton.
Measurement of area of torso burned vs. time after
ignition is another way of indicating the velocity of flame
propagation previously shown in Fig. 4-11. If one records
these data on flame propagation (area of burn) at
specified time intervals, e.g. 30 sec, 60 sec, and

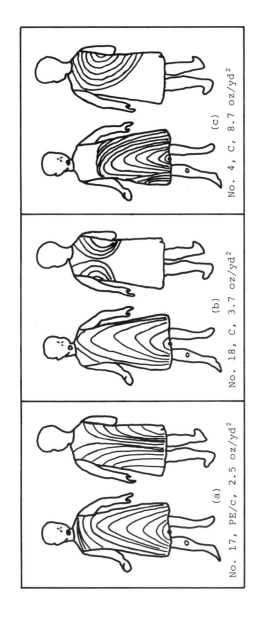

(a) No. 17, PE/c, 2.5 oz/yd^2

(b) No. 18, C, 3.7 oz/yd^2

(c) No. 4, C, 8.7 oz/yd^2

FIGURE 4-10 EXAMPLES OF FLAME SPREAD PATTERNS ON DRESSES IGNITED AT THE FRONT HEM. CONTOURS SHOW POSITIONS OF FLAME BASE AT SELECTED TIMES, THOUGH NOT AT REGULAR INTERVALS. THE RATE OF FLAME SPREAD IS GREATER FOR (A) THAN FOR (B), AND FOR (B) THAN FOR (C). IN GENERAL, AS IN (B) AND (C), THE LOWER PORTION OF THE BACK OF THE DRESS IS EVENTUALLY CUT OFF BY THE FLAMES AND FALLS AWAY [28]

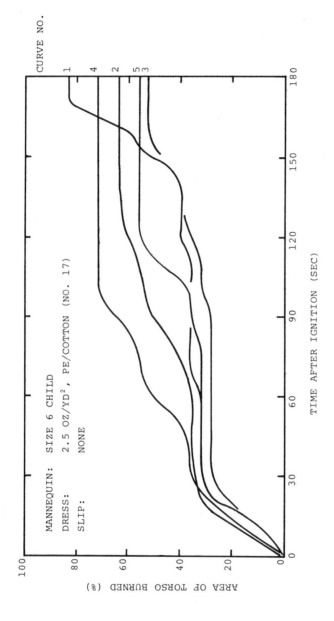

FIGURE 4-11   REPLICATE GARMENT BURNS ON MANNEQUIN (INDIVIDUAL CURVES)   [28]

216

90 seconds after ignition, it becomes possible to characterize the burning behavior of a given fabric in terms of
three data points. The data points obtained for area of
burn after a given time of flame propagation can be plotted
vs. fabric weight and also vs. fiber content. This has
been done by the Gillette group, as shown in Fig. 4-12.
These data indicate the same kind of inverse relationship
between velocity of flame propagation and fabric weight
which was observed in burning of fabric strips in proximity
to a skin simulant. This is shown to exist for each of the
time intervals reported. Further, it is indicated that
at comparable fabric weight, cotton dresses burn more
quickly than their polyester/cotton blend counterparts,
even though it was reported that ultimate areas of burn
are similar for cotton and for polyester/cotton blends.

The same data can be expressed in terms of duration of
burn vs. fabric weight, as shown in Fig. 4-13; it is clear
that at comparable weight, cotton fabrics burn more rapidly
and therefore show a shorter duration of burn than the
polyester/cotton fabrics. It should be noted that the
relative behaviors of 100% cotton and of polyester/cotton
fabrics in the garment tests are consistent with results of
tests with fabric strips over a skin simulant under some
conditions. In particular, it was shown (Fig. 4-6) that
average flame speeds of five cotton fabrics in the 45°
upward burning test for fabric strips were higher than
those of six polyester/cotton fabrics in the same weight
range. However, in horizontal burning tests and in 90°
upward burning tests with fabric strips, polyester/cotton
was reported to burn more rapidly. The feature to be
stressed, of course, is that burning of a garment on a
mannequin involves a wide variety of directional constraints
with varying proximities between fabric and skin, and it is
not surprising that observations on the average flame
propagation speeds in mannequin burn tests do not necessarily
correspond to the results of tests carried out on fabric
strips under closely controlled conditions of orientation

217

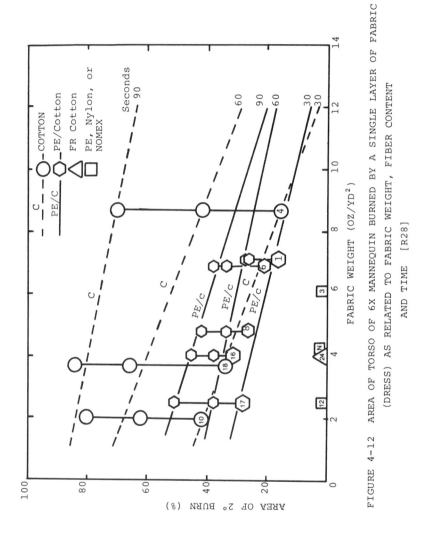

FIGURE 4-12 AREA OF TORSO OF 6X MANNEQUIN BURNED BY A SINGLE LAYER OF FABRIC (DRESS) AS RELATED TO FABRIC WEIGHT, FIBER CONTENT AND TIME [R28]

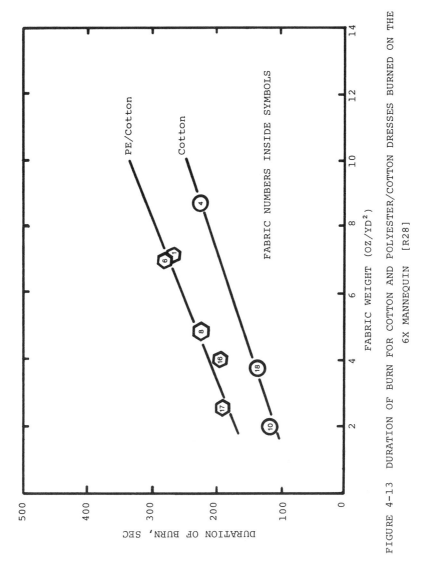

FIGURE 4-13   DURATION OF BURN FOR COTTON AND POLYESTER/COTTON DRESSES BURNED ON THE
6X MANNEQUIN [R28]

and fabric/skin simulant spacing.

Data obtained in mannequin burn tests can also be
reported in terms of the area of torso estimated to have
suffered second-degree burns at the end of the fire test.
This is shown in Fig. 4-14, which includes data for tests
on garment combinations and will be discussed below in
Section 4.3.2.

4.3.2 Garment Combinations and Design

When an undergarment was provided, and dress/slip
combinations were examined in mannequin burn tests, the duration
of burning and the heat transferred to the burned area in the
case of cellulosic fabrics were similar to those observed in
burning single garments made from a fabric of weight comparable
to the combined weight of the dress/slip assembly. However,
this relationship did not necessarily apply when a thermoplastic
fabric or a fire resistant fabric was a component of the garment
combination.

Data for area of burn at the end of the fire on clothed
mannequins are shown in Fig. 4-14. Fabric combinations used
for dress and slip are indicated to include polyester/cotton,
polyester, cotton, nylon, fire-resistant cotton and NOMEX[R]. The
data are so scattered that area of burn cannot be generally
related to the combined weight of fabric in the garments.
Several factors are observed to influence the results. For
example, if a fire-resistant garment is placed on the inside of
the combination, the area of burn is less than for the case where
no inner garment (slip) is used. Examples of this behavior are
observed in the combinations of fabric #16 (a polyester/cotton
blend) with fabric #24 (a fire-resistant cotton fabric). When a
garment made from fabric #16 was burned by itself, area of burn
was just over 80%. When the PE/c fabric #16 was used for the
outergarment with a flame-resistant cotton (#24) for the slip,
burn area was 75%. On the other hand, if the fire-resistant
cotton (#24) was used for the dress, and the polyester/cotton
(#16) for the slip, and the combination was burned together, the
area of burn fell below 70%. A similar trend was noted for the
combination of PE/c fabric #16, the polyester/cotton, with a

220

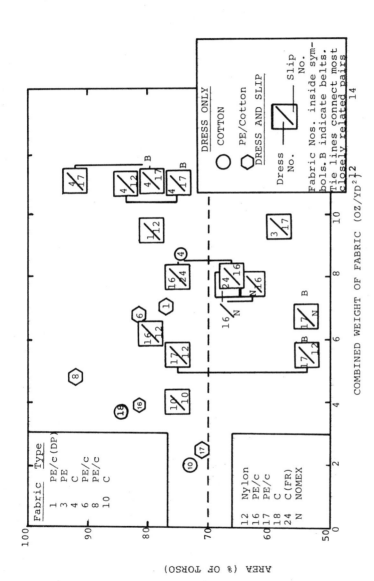

FIGURE 4-14 AREA OF TORSO OF 6X MANNEQUIN CALCULATED TO HAVE 2° BURNS AT END OF FIRE [R28]

221

NOMEX fabric. PE/c fabric #16, as the dress on the outside, with NOMEX as the slip, resulted in about 67% burn area, while the combination of NOMEX for the dress and the PE/c #16 as the slip fabric gave about 64% burn area.

Use of a thermoplastic fabric as a slip material under a cotton or a polyester/cotton dress resulted in an increase in the duration of the burn and, in most cases, an increase in the area of burn as well. Note, for example, fabric #17, a polyester/cotton burned by itself as compared to fabric #17 burned over a nylon slip (fabric #12); note, fabric #16 alone (another polyester/cotton fabric) as against fabric #16 over the nylon fabric #12; note fabric #4 alone (a cotton fabric) as against fabric #4 over the nylon fabric #12; finally, note fabric #1, a polyester/cotton blend treated with durable press alone as against over a nylon (fabric #12) slip.

Other data reported by Gillette on mannequin burn tests suggested that panties provide no significant protection against burn injury, although a bra does provide some protection. Flame propagation in properly fitted slacks ignited at the knee was slower than that for a skirt of the same material ignited at the hem, a result which is not unexpected in view of the different fabric-to-body spacings in the two situations. Suspender-supported slacks with an oversized waist reflected an enhanced rate of flame propagation, again suggesting the importance of fabric-to-body distances, and of chimney effects.

The presence of a belt in the dress design slowed down the rate of flame spread and reduced the area of burn. This is indicated in Fig. 4-14 for some fabric combinations where the reduction in area of burn due to the inclusion of a belt in the dress design was 10%, or higher. For example, in the case of a dress from fabric #17, a polyester/cotton, with fabric #12, a nylon, as a slip material, the reduction in burn area due to a belt was about 20%.

4.3.3 Comparison of Single Garments and Combinations

The effects observed in burning single fabrics as against garment combinations may be summarized as shown in Figs. 4-15 and 4-16. The area of second-degree burns noted

222

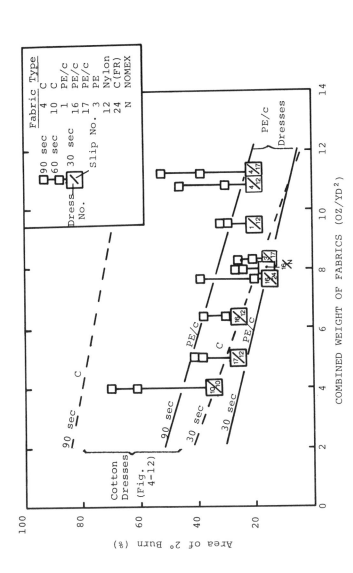

FIGURE 4-15 AREA OF TORSO OF 6X MANNEQUIN BURNED BY DRESS-SLIP COMBINATION AS
COMPARED TO DRESSES ALONE (DATA FROM FIG. 4-12) FOR CLARITY, SOME SETS OF POINTS
HAVE BEEN SLIGHTLY DISPLACED ALONG THE ABSCISSA [R28]

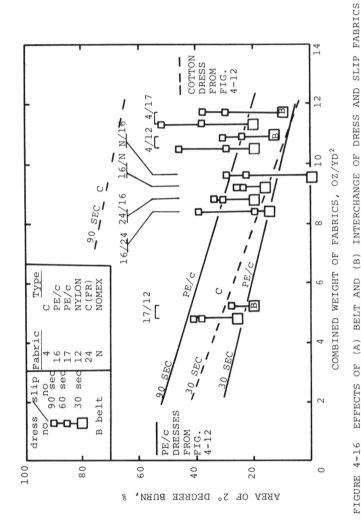

FIGURE 4-16   EFFECTS OF (A) BELT AND (B) INTERCHANGE OF DRESS AND SLIP FABRICS
ON THE AREA OF THE TORSO OF THE 6X MANNEQUIN BURNED BY DRESS AND
SLIP COMBINATIONS. FOR CLARITY, SOME SET OF POINTS HAVE BEEN
SLIGHTLY DISPLACED ALONG THE ABSCISSA   [R28]

30 sec., 60 sec. and 90 sec. following ignition is plotted
against the combined weight of fabrics used in the slip and
dress assembly. The figure also shows lines for data obtained
at 30 sec. and at 90 sec. in burn tests on dresses only made
from cotton or polyester/cotton blend fabrics (previously presented in Fig. 4-12). The Gillette group concludes that "as
a rough generalization it appears that the combinations fall
into two categories--those which behave like polyester/cotton
blends and show an increase of 15-25% in the area burned between
30 and 90 sec. and those which behave more like cotton and show
an increase of 25-50% in the 30- to 90-sec. interval. Dress/
slip combinations of polyester/cotton blends and thermoplastic
fabrics (17/12, 16/12, 16/n, 3/17 and 1/12, where the dress
fabric number is always given first) appear to behave roughly
like the blend dresses whose fabric weights were the sums of
those of the individual garments. On the other hand, all
cotton combinations (10/10), combinations of cotton and blends
(4/17), or cotton and thermoplastics (4/12) approach more
closely the behavior of all cotton fabrics of the summed dress
and slip fabric weights". Exceptions to this generalization are
noted.

The data summarized in Fig. 4-15 suggest that the
dominant effect on flame-propagation velocity for a wide range
of commercial textile materials is the combined weight of
fabrics contained in close proximity garments (referring to
dresses worn over slips in contrast to dresses worn over
panties). The data further suggest that for many material
combinations of cellulosics and thermoplastics the fiber content
is not the critical element, but rather the weight of the
combination determines the rate of flame propagation, provided
that sufficient cellulosic fiber to sustain combustion is included in the assembly. In other words, while it has been
recognized that cellulosic-thermoplastic blend fabrics are
flammable materials, it must also be stressed that cellulosic
and thermoplastic fiber garment combinations are likewise
flammable. Their behavior, as long as combustion is sustained,

reveals a strong relationship between velocity of flame propagation and mass-per-unit area of the combined textile system.

4.3.4 Observations on the "Ease of Extinguishment"

In the garment tests on mannequins, the fire was extinguished after the specified time by beating it out with asbestos gloves.

Comments on the ease of extinguishing garments have been reported in the Gillette study, as follows: "At comparable states of flame advancement, the lighter weight fabrics are easier to extinguish, whereas blend or thermoplastic garments complicate the extinguishing process by tending to stick to the gloves. The few cases in which garments could not be safely extinguished always involved well-established fires of dresses or slacks of cotton or of polyester/cotton blends of 6 oz/sq. yd., or heavier fabrics. Extinction is easiest when the garment/mannequin distance is low. The extinction is difficult in convoluted areas such as the crotch in burns of slacks, or the armpits".

4.4 COMPARISON OF FLAME SPREAD RESULTS ON FABRIC STRIPS AND ON GARMENTS

The M.I.T. Fuels Research Laboratory reported results for the average flame speed in selected GIRCFF fabrics of various weights. Two series of fabrics were studied, namely 100% cotton fabrics and polyester/cotton blends, in various orientations (0°, 45° and 90° to the horizontal) and at various spacings between the burning fabric and the skin simulant. The results indicate the effect of such variables as fiber composition, fabric weight, angle of fabric during the burn test, as well as direction of the burn (up or down) and spacing between fabric and skin simulant. The effects of these variables can also be seen in the data reported by Gillette Research Institute. For example, in Fig. 4-10, it was shown that the most rapid burning of the fabrics takes place during vertical upward burning of that portion of the dress which is not in contact with the body. It was shown that burning to the side is a much slower process than burning upward. The data also indicated that the rate of

flame propagation is reduced significantly as the flames reach
the area where the garment is in close contact with the body--
at the waist, or at the chest. The major effects of fabric
weight, orientation and fabric/body spacing observed in the
fabric strip tests were also observed in the garment tests.
And trends noted in the two series of tests are in good agreement.

4.5 EFFECTS OF FABRIC PARAMETERS (NOT CONSIDERED BY GIRCFF
     PROGRAM) ON FLAME SPREADING

It has often been suggested that air-permeability of fabrics
can influence flammability, but results of flame propagation
tests on a variety of commercial materials have shown that air
permeability is not a significant factor with regard to the
average speed of flame propagation. The data recorded for
flame speed by the Gillette group and by the Fuels Research
Laboratory at M.I.T. have shown that the wide variations in
air permeability for the GIRCFF fabric series did not affect
the clear relationship between rate of flame propagation and
weight-per-unit area. The lack of dependence of speed of
flame spreading on air permeability is shown graphically in Fig.
4-17. These data were obtained in the Department of Mechanical
Engineering at M.I.T.[31] for a series of cellulosic materials
(including fabrics) of different air permeabilities. They show
results of horizontal burning tests and vertically downward
burning tests for cellulosic materials ranging from less than
2 oz/yd^2 up to 20 oz/yd^2, with air permeabilities ranging from
well below 100 ft^3/ft^2/min up to 600 ft^3/ft^2/min. Data previously reported by Bernskjold[12] for rate of flame propagation of cellulosic fabrics in an upward burning test are
shown in Fig. 4-18, and show the mean porosity of the fabrics,
calculated from measurements of weight-per-unit area,
thickness, and fabric density. No significant effect of mean
porosity is apparent over the range of 0.61 to 0.97 from the
data shown in Fig. 4-18. The data of Fig. 4-18 were replotted
by Bernskjold to show the relationship of flame propagation
speed to the reciprocal of weight-per-unit area corresponding
to a linear relationship. This is similar to the log-log

227

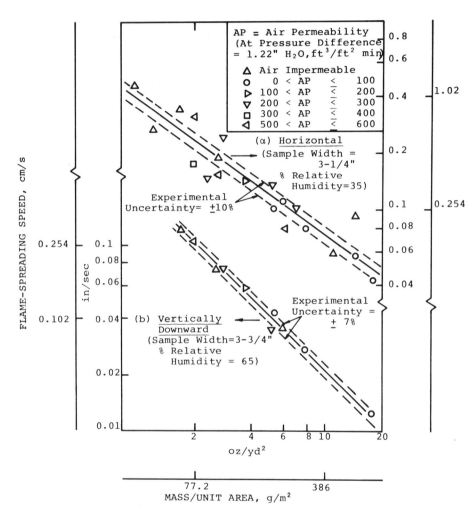

FIGURE 4-17 FLAME-SPREADING SPEED VS. MASS/UNIT AREA FOR
CELLULOSIC MATERIALS OF DIFFERENT AIR PERMEABILITIES. (RIGHT
AND LEFT ORDINATES ARE FOR HORIZONTAL AND FOR VERTICALLY
DOWNWARD ORIENTATION RESPECTIVELY) [31]

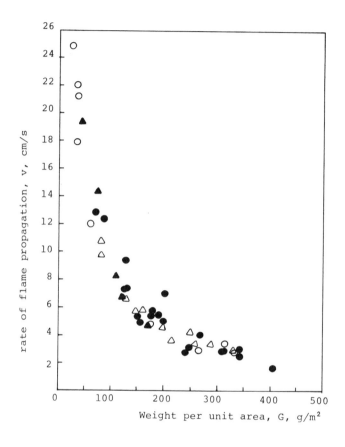

O	mean porosity $0.90 < \bar{e} < 0.975$
●	mean porosity $0.80 < \bar{e} < 0.90$
△	mean porosity $0.70 < \bar{e} < 0.80$
▲	mean porosity $0.61 < \bar{e} < 0.70$

FIGURE 4-18   RATE OF VERTICALLY UPWARD FLAME PROPAGATION
AS A FUNCTION OF THE WEIGHT PER UNIT AREA   [12]

plot of Fig. 4-17, except that the conditions of burning differ. The Bernskjold data were obtained in vertical upward burning, while the data in Fig. 4-17 correspond to horizontal (a) and vertical downward (b) burning. The graph of Fig. 4-19 was obtained by replotting some of Bernskjold's data for woven fabrics, knit fabrics, non-woven fabrics, paper and needle felts, and the lack of dependence of flame propagation rates on structural parameters other than mass-per-unit area is again demonstrated.

The apparent lack of effect of fabric structure on flame spreading over cellulosic materials within the range discussed can be explained[31] as follows: the physical dimension associated with the textile structure is of the order of the diameter of the yarns, or of any spacing between them, that is, of the order of 10^{-1} to 10^{-2} in. On the other hand, the physical dimension associated with flame spreading is the pyrolysis length and is of the order of an inch, depending on the fabric mass-per-unit area. (This pyrolysis length represents the effective diameter of a hypothetical burner moving with the flame.) Thus, the differences in structure are on a much smaller scale and cannot affect the flame spreading speed, which depends on processes occurring over the larger area of the pyrolysis region.

If differences in structure are comparable to the size of the flame, then they have an effect on the flame speed. To illustrate this point, experiments were run on rayon rovings placed side by side with their axis in a horizontal plane. It was found[31] that the rate of flame spreading depends on the spacing between the two rovings. Two effects have been observed. First, as suggested by Toong[31a], when the spacing is decreased, the induced air flow between the rovings is increased and, consequently, their flame size and burning rates increase. At very small spacings, however, the two flames merge into one, and the burning rate is reduced due to oxygen depletion in the air flow between the rovings. Second, as the spacing is decreased, stronger induced air flows are

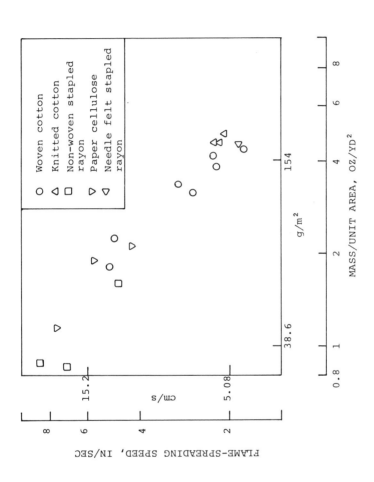

FIGURE 4-19 VERTICALLY UPWARD FLAME-SPREADING SPEED VERSUS MASS/UNIT AREA FOR MATERIALS OF DIFFERENT CONSTRUCTION (FROM BERNSKIOLD[12], TABLES 6.1 and 6.2) VIA [31]

set up underneath the burning elements. These currents enhance the convective forward heat transfer under the rovings and, consequently, the flame spreading speed. These two effects are competing and the flame spreading speed is found to be at a maximum at a spacing-roving diameter of the order of one.

The above results and discussions apply for cellulosic materials. However, for thermoplastic fabrics, Miller and Goswami at TRI reported[35] large effects on flame spreading speed due to small variations in structure. For example, they measured 100% change in flame speeds between burning along and against the twill direction of a polyester fabric. In this case, small variations in structure and in interlacing geometries can have significant effects on the amounts of shrinkage of the polyester fabric, when subjected to a heat source. The physical dimension of shrinkage is of the same order of magnitude of the pyrolysis region (and sometimes even larger) and could conceivably affect the flame speed. Thus, contrary to the case of cellulosic materials, small variations in the structure of thermoplastics could (through their effects on shrinkage) affect the flame speed.

4.6 OVERVIEW

In this chapter, the results of the M.I.T. Fuels Research Laboratory study of flame propagation in the GIRCFF fabrics and of the Gillette Research Institute study of flame propagation in single garments and garment combinations have been discussed in detail. So much data from the M.I.T. and Gillette studies have been incorporated in this report that it is desirable to summarize some of the principal conclusions in general terms and to compare them with information from the literature.

Comparisons of flame spreading in the GIRCFF fabrics burned alone (single layer of fabric) have shown that the phenomena of flame propagation are affected in major ways by four variables:

Fiber composition causes gross differences in burning behavior, primarily with respect to the physical response of the fabric after ignition. Thermoplastic fabrics melt, shrink, generally do not propagate the flame, and are not amenable to quantitative measurements of flame propagation speeds. Non-thermoplastic fabrics continue to burn after ignition. The effect of fiber composition on the rate of flame propagation in non-thermoplastics (e.g. cotton, polyester/cotton and polyester/rayon) is indicated to be relatively small, and far less important than other variables (see below). In the case of flame resistant fabrics (e.g. cotton treated with fire retardants) flame propagation is inhibited and the above generalizations would not apply.

Fabric orientation has a great effect on the rate of flame propagation. For burning of a given fabric, maximum flame speed is noted for upward burning at 90° (vertical) orientation. For upward burning at 45° orientation, the average flame speed is slightly lower, and it is much lower for horizontal burning. These relationships observed on fabric strips have their counterpart in garment tests: maximum flame speed occurs for hem ignition of a dress (corresponding to 90° upward burning) and lateral flame spread in a garment is much slower than upward propagation.

Fabric weight-per-unit area has a dominant effect on flame propagation speed in all fabrics and garments where burning is sustained (non-thermoplastic). For all test geometries and fabrics investigated, the inverse relationship of fabric weight to flame speed has been confirmed.

Spacing--defined as the distance of the fabric from the body (or skin simulant)--affects flame velocity in fabric strip tests, where it can be controlled, also in garment tests where it is grossly and non-uniformly dependent on garment design and fit.

For fabrics burned in combinations either as double layer composite strips, or as garment assemblies, the complexity of the system is such that few generalizations are warranted. Interactions of thermoplastic and non-thermoplastic fabrics in fabric strip tests and in garment tests, and the multiplicity of orientations and spacings in a garment assembly (e.g.

233

dress/slip or dress/panties) preclude the possibility of generalizations regarding these effects (fiber composition, orientation and spacing) on flame spreading in combinations. However, the important effect of weight can be seen in fabric combinations consisting of non-thermoplastic components, where the rate of flame propagation can be related to the combined weight of the fabric components.

The findings summarized above are generally consistent with data on flame propagation reported in the literature, and they advance our knowledge in several ways. The previously proposed dependence of flame speed on fabric weight, and on orientation, has been confirmed and extended to include additional fabric types and experimental conditions. The effect of spacing on flame propagation rate has been defined and supported by experimental data. Also, a measure of correlation has been established for flame propagation tests carried out with the same fabrics and fabric combinations on fabric strips and on garments.

CHAPTER V

BURN INJURY

5.1   INTRODUCTION   236
      5.1.1   Description of the skin   238

5.2   BURN INJURY CRITERIA   241
      5.2.1   Studies of Henriques and Stoll   241
      5.2.2   Energy Density Criterion   254
      5.2.3   M.I.T.'s Modification of Henrique's Burn
              Injury Criteria   258
      5.2.4   Summary   260

5.3   PREVIOUS STUDIES ON BURN INJURY POTENTIAL OF FABRICS   265
      5.3.1   Fabric Strip Studies   265
      5.3.2   Mannequin Studies   269

5.4   BURN INJURY POTENTIAL OF GIRCFF FABRICS IN
      GARMENT FORM   273
      5.4.1   Overall Heat Transfer from Burning Single
              Garments   273
      5.4.2   Local Heat Transfer from Burning Single
              Garments   279
      5.4.3   Fabric Combinations and Garment Designs   283

5.5   BURN INJURY POTENTIAL OF GIRCFF FABRICS IN STRIP
      FORM   287
      5.5.1   Skin Simulants   288
      5.5.2   Effects of Fabric-Skin Spacing and Gravita-
              tional Orientation   290
      5.5.3   Horizontal Strips   294
      5.5.4   Inclined (45°) and Vertical Strips   297
      5.5.5   Fabric Combinations   299

5.6   MECHANISMS OF HEAT TRANSFER TO SKIN   306
      5.6.1   Horizontal Samples   309
      5.6.2   Vertical Samples   314
      5.6.3   Effects of Fabric-Skin Spacing   319
      5.6.4   Summary   324

5.7   OVERVIEW OF STUDIES OF BURN INJURIES   326

CHAPTER V

BURN INJURY

5.1 INTRODUCTION

In the preceding chapter, we have considered the factors
which influence rate of flame propagation in a burning fabric
sample, or in a burning garment. It was indicated that fabric
weight is a dominant factor influencing the rate of flame
propagation in cellulosic fabrics; within the range of fabric
structures examined, cloth porosity, or air permeability is not
an important variable. It was stressed that the circumstances
of the burning, such as the orientation of the fabric with
respect to the horizontal or whether the flame is moving upward
or downward, also influence the rate of flame propagation. In
addition, the proximity of the skin to the burning fabric serves
to modify the size of the flame and the speed of its propagation.

Any surface which serves to quench a flame by acting as a
heat sink is, of course, subject to temperature increases and,
consequently, to changes in physical and chemical structure.
The current chapter is concerned with the changes which take place
in the human skin as a result of rapid heat transfer from an
adjacent burning fabric, i.e. the phenomenon of burn injury.

Heat transfer to and from the surface of the human body has
been a subject of study by physiologists and textile technologists for a number of years. A major expansion of these studies
was motivated by the pressing need of protecting men under
extreme conditions of heat and cold during World War II.
Subsequently, with the development of atomic weapons, considerable
attention was devoted to the development of thermally protective
fabrics and to the study of their behavior when subjected to
extremely high levels of radiative flux. Much work also con-

cerned the development of flameproof protective clothing for fuel handlers and aircraft personnel.

More recently, there has been a spate of books on the subject of human physiology and clothing science. One major work was derived from U.S. Army studies [36], a second came from a physiological research center of the British Ministry of Defense [37] and the third from the Technical University of Denmark [38]. While the above references focus on human physiology as it relates to comfort, other reports concentrate on human physiology as relating to discomfort, or more specifically, to pain sensations and human injury. A classic reference in this area is the work of Hardy and his co-workers [39]. More recently, Stoll has studied condition extremes which lead to human discomfort and injury [40], directing attention to the overall phenomenon of heat exchange between the human body and the total environment, as well as to the effect of protective garment-barriers on that exchange. Stoll also focused on the human enclosure which forms an interface with the environment - the skin.

Stoll points out that it is "the skin rather than the body as a whole, which, by virtue of its many receptors, determines the physical characteristics of an environment which is interpreted as comfortable. In addition to this role as a detector, the skin functions thermally as a heat generator, absorber, transmitter, radiator, conductor and vaporizer. It is not often appreciated as a separate organ of the body, as vital to life as the heart, yet if 40% of the skin is destroyed, the life of the entire individual is severely jeopardized, while death is inevitable on destruction of the total skin area. If, however, the individual survives extensive dermal trauma, the skin is remarkably regenerative and recovers, though frequently requiring the assistance of grafts."

5.1.1 Description of the Skin

"The complexity of this organ may be illustrated to some
extent in Fig. 5-1. It is seen that the skin is a composite
of morphologically distinct tissues. Within these layers of
tissue are situated many discrete structures, including hair
follicles, sweat glands, sensory receptors, capillary blood
vessels, and an extensive network of fine nerves which contact
these structures and relay signals to and from the central
nervous systems. The whole skin is comprised of two main
portions - the epidermis and the dermis. The epidermis is subdivided into four layers, the stratum corneum, stratum lucidum,
stratum granulosum, and stratum germinativum. One or two of
these layers (viz., stratum lucidum and stratum granulosum) may
be absent in some areas of the body. The dermis is subdivided
into two layers, subepithelial or papillary layer and the
reticular layer, both of which are composed of connective
tissue. All layers except the stratum corneum are living,
growing tissue and therefore produce metabolic heat." [40]

It has also been pointed out that the skin comprises 6%
of the total body weight, that it has the largest area of any
organ in the body, and possesses a very limited thickness,
which may be more than 5mm. on the back and as little as 0.5mm
on the eyelids (an average thickness may be taken as 1-2mm).

Clearly, the initial effect of heat transfer from a burning fabric to human body is to raise the skin temperature. It
is of interest to note the complicated results incurred when
the skin tissue is raised, or lowered, to specific temperatures.
Table 5-1, prepared by Stoll, illustrates the range of
sensations, skin color and injury which may be expected when
the tissue temperature is established anywhere from 25° F. to
162°F., ranging from freezing to severe burning and irreversible burn damage.

238

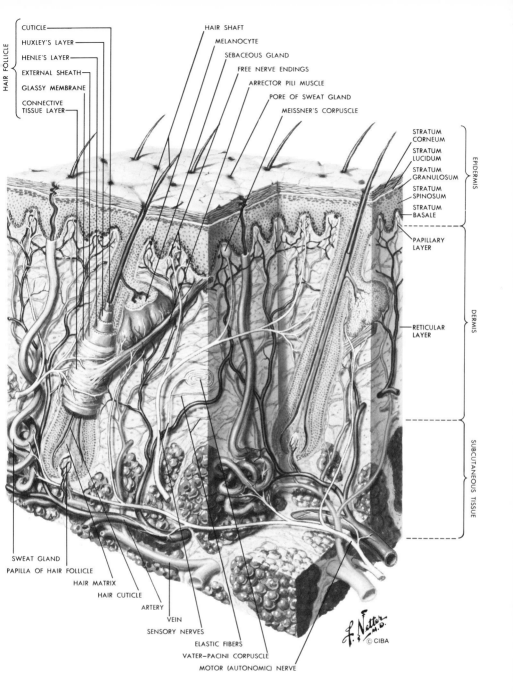

HAIR FOLLICLE

CUTICLE
HUXLEY'S LAYER
HENLE'S LAYER
EXTERNAL SHEATH
GLASSY MEMBRANE
CONNECTIVE TISSUE LAYER

HAIR SHAFT
MELANOCYTE
SEBACEOUS GLAND
FREE NERVE ENDINGS
ARRECTOR PILI MUSCLE
PORE OF SWEAT GLAND
MEISSNER'S CORPUSCLE

STRATUM CORNEUM
STRATUM LUCIDUM
STRATUM GRANULOSUM
STRATUM SPINOSUM
STRATUM BASALE
EPIDERMIS

PAPILLARY LAYER

RETICULAR LAYER
DERMIS

SUBCUTANEOUS TISSUE

SWEAT GLAND
PAPILLA OF HAIR FOLLICLE
HAIR MATRIX
HAIR CUTICLE
ARTERY
VEIN
SENSORY NERVES
ELASTIC FIBERS
VATER-PACINI CORPUSCLE
MOTOR (AUTONOMIC) NERVE

FIGURE 5-1 SECTION OF HUMAN SKIN, AFTER CIBA

TABLE 5-1

THERMAL SENSATIONS AND ASSOCIATED EFFECTS THROUGHOUT RANGE OF TEMPERATURES COMPATIBLE WITH TISSUE LIFE [40]

Sensation	Skin color	Tissue temperature °C	Tissue temperature °F	Process	Injury
Numbness	White			Protein coagulation	Irreversible
		72	162		
	Mottled red and white	68		Thermal inactivation of tissue constituents	Possibly reversible
		64			
		60	140		
Maximum pain	Bright red	56			
Severe pain		52			Reversible
	Light red	48			
Threshold pain		44	111		
Hot					
	Flushed	40			
Warm		36		Normal metabolism	None
			93		
Neutral	Flesh	32			
		28	82		
Cool					
	Blanched	24			
Cold	Red	20			
			64		
Threshold pain	Bluish red	16		Physicochemical inactivation of tissue constituents	Reversible
Severe pain	Reddish purple	12			
			50		
		8			Possibly reversible
	Bright pink	4			
		0	32		
	White	-4	25	Protein coagulation	Irreversible
Numbness					

240

The processes of heating a fabric to ignition and heating the skin to burn injury manifest some roughly parallel features. If one knows the heating intensity and the efficiency of heat transfer, as well as the thermal properties of the receiver, one may calculate the time of exposure needed to raise the temperature of the fabric to the ignition point or the temperature of the skin tissue to a level where temporary or permanent burn damage will occur.

Stoll reports on the thermal properties and optical properties of skin in Table 5-2. Here we see many of the parameters which have been measured and reported by the Georgia Tech group studying the phenomenon of heat transfer from an ignition source to a textile fabric. The one property listed for skin, but not included in the data for textile fabrics, is obviously the heat production of metabolism of the dermis, listed as 240 kcal/day. Stoll emphasizes that, "although these quantities are frequently referred to as constants, it is obvious that they must be far more variable than constant since the medium itself is not only heterogeneous, but also of varying thickness and composition from place to place on the body surface. Furthermore, because of the presence of thermal and other sensory detectors of the skin, its thermal properties are influenced by an efficient feedback system operating through cutaneous blood flow system, as well as through a voluntary or cortical control whereby the entire organism is activated to remove itself from hazardous environments".

5.2 BURN INJURY CRITERIA

5.2.1 Studies of Henriques and Stoll

One may note in Table 5-1 thermal sensations experienced going from warm to hot and then to threshold pain as the

241

TABLE 5-2

THERMAL PROPERTIES AND OPTICAL PROPERTIES OF SKIN [40]

Property	Value
Heat production	240 kcal/day
Conductance	9-30 kcal/m^{2}hr °C
Thermal conductivity (k)	(1.5±0.3) x 10^{-3}cal/cm sec °C at 23-25 °C ambient
Diffusivity (k/ρc)	7 x 10^{-4} cm^2/sec (surface layer 0.26 mm thick)
Thermal inertia (kρc)	90-400 x 10^{-5} cal^2/cm^4 sec °C^2
Heat capacity (c)	0.8 cal/gm °C
Emissivity (infrared)	0.99
Reflectance (wavelength dependent)	Maxima 0.6 to 1.1 μ Minima < 0.3 and > 1.2 μ
Transmittance (wavelength dependent)	Maxima 1.2, 1.7, 2.2, 6, 11 μ Minima 0.5, 1.4, 1.9, 3, 7, 12 μ

tissue temperature goes from 36° C. to 40° and then to 44° and above. It is of interest to note some of the early sources of such observations. For example, Moritz and Henriques [41, 42, 43] working at Harvard Medical School in the late forties, undertook experiments where they applied hot running water directly to skin surfaces for various times and observed visually and microscopically, the nature of burn damage encountered. They reported that when the water temperature was below 44° the rate of cellular reparative processes seemed to offset the rate of the thermal damage processes. As a result, second-degree burns were not observed even after six hours of exposure to the hot running water. A moderate second-degree burn corresponds to separation of the 80-100 μm epidermis from the underlying dermis, i.e., the formation of a blister. When the water temperature was raised above 44° the damage rate was nearly doubled for each degree rise in temperature up to 51°. Still within this range of 44° to 51°, the exposure times required for burn damage were quite large and the epidermal layer reached steady state temperature well before any burn damage was observed. At water temperature above 51°, the time required for burn injury was quite short, occurring before the transepidermal layer reached a steady state value of temperature.

In other words, when the body was exposed to water temperatures between 44° and 51° C., the exposure required for burn injury were large enough so the epidermal basal layer reached a steady state value almost equal to the surface temperature of the skin. In contrast, at temperatures above 51°, the exposure times for burn injury were quite short and transient conditions persisted, meaning that surface temperatures were considerably different than temperatures below the skin surface at a depth of 80 microns, corresponding to the junction between the epidermis and the dermis. Since second-degree burn damage was characterized by epidermal necrosis down to the 80 μm.

level, it followed that knowledge of the temperature-time relationship at this basal layer of the skin was necessary for monitoring of progressive burn injury.

As was shown subsequently [40] the rate at which skin tissue damage occurs is exponentially related to the skin temperature as shown in Fig. 5-2. The rate of damage at 50° C. is one hundred times that occurring at 45° C. and there appears to be a critical temperature at 50° C. at which point the slope is reduced somewhat above 50°. It is pointed out [40] that at higher temperatures the damage occurs very rapidly, with" instantaneous"destruction of the skin occurring at about 72°C.

Since it was virtually impossible to take measurements of the temperature at the junction between the epidermis and the dermis, Henriques [44, 45] calculated the temperature at the basal epidermal layer using transient heat conduction solutions for heat transfer through an opaque, semi-infinite solid. He assumed that rate of injury at the base of the epidermal layer was a rate process, depending strongly on temperature of the tissue at that level and formulated the damage integral Ω where

$$\Omega = \int_0^{t_e} \frac{d\Omega}{dt} \, dt = \int_0^{t_e} P \exp\left(\frac{-\Delta E}{RT_t}\right) dt \, . \qquad (5.1)$$

Here T_t is the absolute temperature of the tissue 80 microns below the surface (at the basal layer of the epidermis) at time t, t_e is the time of exposure, R is the universal gas constant, P is the frequency factor, ΔE is the activation energy, and Ω is the degree of tissue injury, or damage.

The damage integral Ω was determined by histological examination of damaged tissue and was assigned the value one at first occurrence of complete tissue death (necrosis). Experiments were then run at relatively low water temperatures (in contact with pigskin)

244

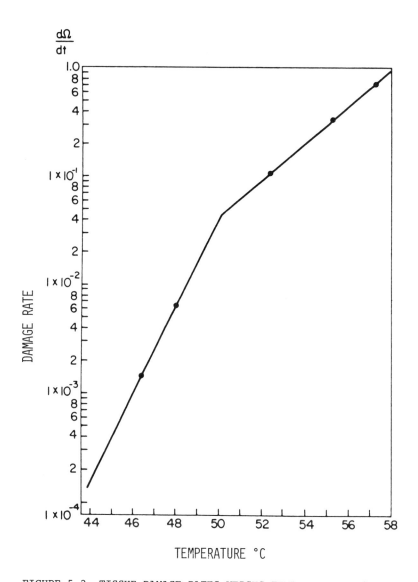

FIGURE 5-2   TISSUE DAMAGE RATES VERSUS TEMPERATURES   [40]

245

so that the temperature at the 80 micron level was essentially
constant and equalled the temperature of the water. Thus, T_t
in (5.1) became independent of the time t, simplifying integration
of the equation. Results of such low-temperature exposures of
pigskin permitted determination of the constants P and E. Using
these constants, it was determined that the value of $\Omega=0.53$,
corresponded to the maximum exposure which could be suffered
without the occurrence of irreversible tissue damage.

When higher test temperatures resulted in shorter exposures
for occurrence of injuries, the temperature of the epidermal basal
layer was no longer equal to the skin surface temperature, and
had to be calculated through use of the heat transfer model
mentioned above. Numerical methods were then used to obtain
values for the damage integral, Ω, based upon equation (5.1).
And a good correlation of the calculated values of Ω with actual
burn damage was reported.

Henriques' approach to the prediction of skin burn damage
has been widely accepted, however, with some major modifications.
Stoll [40] pointed out that tissue damage occurs at all temperatures
above 44°, regardless of whether the temperature is rising or
falling. This means that in the determination of the Henriques
integral, one must integrate over the entire exposure period
during which the subepidermal temperature (at 80 microns depth)
is above 44°, that is, during the heating up and during the cooling
portion of the cycle. Experiments to illustrate this fact are
reported in Figure 5-3, which shows time-temperature histories
during and following thermal irradiation at two different levels
of radiation intensity. In Figure 5-4 two cases are reported
where equal damage is produced (at level $\Omega = 1$) but different
proportions of that damage occur during the cooling portion of
the cycle, depending upon whether the tests were conducted at
a low level or a high level of irradiance. At the lower level
only 10% of the damage occurs in the cooling cycle, whereas at

246

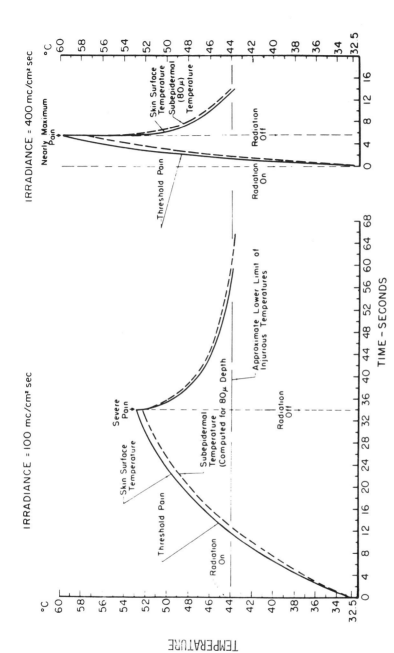

FIGURE 5-3 TEMPERATURE-TIME HISTORIES DURING AND FOLLOWING THERMAL IRRADIATION
AT TWO LEVELS OF INTENSITY [40]

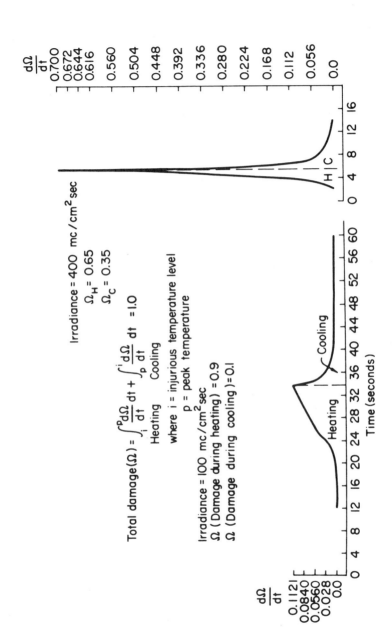

FIGURE 5-4 DAMAGE DURING HEATING AND COOLING [40]

the higher level exposure, more than a third of the damage occurs during cooling. The relative contribution of cooling period damage to total damage is indicated in Fig. 5-5 as a function of heating rate.

Stoll[46] has further shown that the temperature rise in skin and the tolerance time to a given damage level is dependent on the heat flux--and this relationship is much the same regardless of the mode of heat transfer. Figure 5-6 shows a plot of absorption rate vs. exposure time to achieve white burns on anaesthetized rats both for convection (flame) and for radiation heating.

Similar curves have been drawn relating energy absorption rate vs. time of human skin exposure at pain level, threshold blister level and full blister level[47]. Such data can be plotted on a log-log scale for a rectangular heat pulse as in Fig. 5-7[46]. The lower line represents the pain threshold while the uppermost line is the blister line. The line marked "survival" plots the condition where pain is felt, but skin damage is reversible without blistering. Stoll transposed these absorption rates to read instead as the temperature rise in a known skin simulant measured after an arbitrary 3-second exposure to a rectangular heat pulse, as shown in Fig. 5-8. These relationships are then used by Stoll to convert maximum skin simulant temperatures measured on the body of a clothed mannequin subjected to a rectangular heat pulse for a given exposure time into a designation of burn severity. While refinements could be made by structuring Fig. 5-8 for first, second, and third degree burns, it was found sufficient in both flame tests and high intensity thermal radiation tests to designate various areas of the mannequin as severely burned (second degree, or worse) or not burned. The overall degree of burn damage was then expressed as the percentage of the exposed surface area which suffered "severe burn" damage.

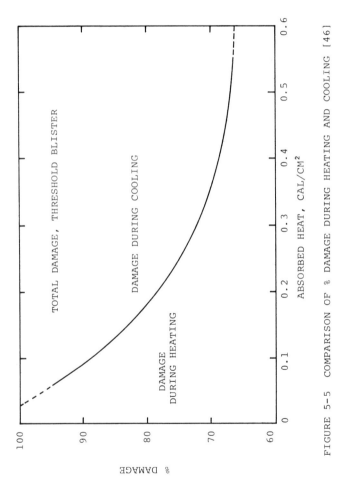

FIGURE 5-5 COMPARISON OF % DAMAGE DURING HEATING AND COOLING [46]

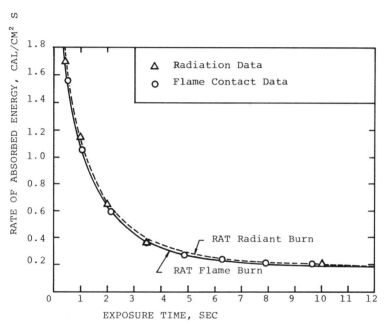

FIGURE 5-6   COMPARISON OF STRENGTH-DURATION CURVES
FOR WHITE BURNS PRODUCED BY FLAME CONTACT AND BY
THERMAL RADIATION [46]

251

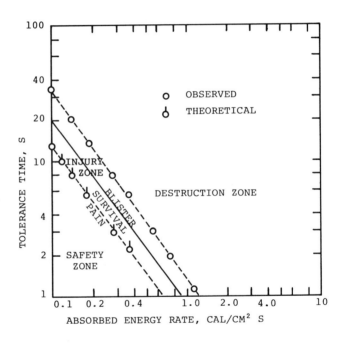

FIGURE 5-7   HUMAN SKIN TOLERANCE TIME VS. RATE OF
ABSORBED THERMAL ENERGY DELIVERED IN A RECTANGULAR
HEAT PULSE [46]

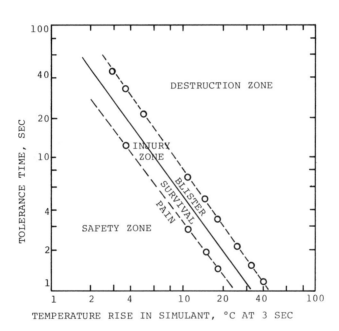

FIGURE 5-8 HUMAN SKIN TOLERANCE TIME INDICATED BY
THE TEMPERATURE RISE MEASURED IN A SKIN SIMULANT AT
3 SECONDS' EXPOSURE TO A RECTANGULAR HEAT PULSE

5.2.2 Energy Density Criterion

Chouinard et al[34], analyzed Stoll's data for absorbed
heat flux vs. tolerance time to occurrence of blisters and
noted that the energy per unit area (i.e. product of absorbed
heat flux and time) required to blister, decreases with increasing
flux. But he noted that for fluxes in the range of 0.3 to 1.0
cal/cm^2 sec which are observed for many fast burning fabrics,
the energy density of 2 cal/cm^2 serves as a threshold level and
heat transfer in excess of this amount will cause epidermal
damage. "Once the epidermal limit has been reached, additional
heat transfer can result in destruction of deeper lying structures. Measurement of the heat-flux, as a function of time, can
thus be related to the severity of burn trauma at any point."

Accordingly, Chouinard et al, used small copper disk
sensors (with thermocouples embedded at their back) and mounted
them flush in the surface of a test mannequin so as to record
heat-flux history at 24 locations during a burn test. From
these histories he determined the number of points at which the
2 cal/cm^2 energy limit was exceeded, and also the average
energy density transferred to these points, the former indicating
the area of severe burning, and the latter indicating the
extent of destruction of deeper tissue over that area.

A similar use of the Stoll criterion for occurrence of a
second degree burn was reported by Arnold et al, at the
Gillette Research Institute[48] who noted "that burn damage to
humans is only caused by heat input at rates in excess of
about 0.05 cal/cm^2 sec and that it takes about 2.0 cal/cm^2
input above this minimum rate for a second degree burn to be
obtained. By determining the number of sensors which had been
subjected to a heat input history sufficient to cause a second
degree burn according to the Stoll criterion, the percent of
the area of the torso which would have suffered second degree,
or deeper, burns in a real life garment fire could be calculated".

To provide an index of severity of burn, as opposed to area of burn, one method was to determine the total heat received per unit area for each sensor by the time the flames had died out, and to calculate the average of these values for the sensors which had exceeded the second degree limit. This gave the average heat transfer to the burned areas. Subsequently, it was possible to use results obtained from the program supported by Cotton Incorporated[48] to correlate heat input with burn depth (see Fig. 5-9a). The significance of Fig. 5-9a is that it suggests that burn depth on the skin of shaved, anaesthetized white rats which have been exposed to burning fabrics increases roughly linearly with heat input from about zero at 5 cal/cm² to 2000μm (2 mm) at about 12 cal/cm². Thus, assuming human skin to have a similar sensitivity to heat, the line in Fig. 5-9a can be used to provide an estimate of the burn depth from the measured heat input. On this basis, a burn depth of 100μm (the approximate thickness of the epidermis) would require slightly more than 5 cal/cm². In using the 5 cal/cm² burn criterion, one must keep in mind the extent of data scatter in Fig. 5-9a.

It is of interest to note that the view that the severity of burn is dependent on the amount of thermal energy entering the skin above a critical temperature was held by Chen[29], who did not consider thermal injury caused by short term exposure to be a rate-controlled process. Chen's criteria were completely empirical and correlated only with the occurrence of second degree mild burns in pigs. As one of its tasks in the GIRCFF program, the Fuels Research Laboratory at M.I.T. undertook to modify and improve various injury criteria recorded in the literature, including Chen's critical energy-critical temperature criteria and Henriques' damage integral criteria.

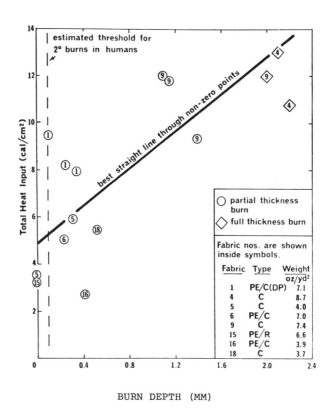

BURN DEPTH (MM)

FIGURE 5-9a   BURN DEPTH OF WHITE RAT SKIN AS A FUNCTION OF
TOTAL HEAT INPUT FROM FABRICS BURNING VERTICALLY UPWARDS
AT 1.5 INCHES FROM THE SKIN, (FULL THICKNESS BURNS MAY GIVE
DEPTHS LOWER THAN WOULD BE OBSERVED FOR THICKER SKIN)   [28]

According to the critical energy-critical temperature criteria, thermal injury of the skin occurs with an increase in its energy above a critical temperature. Chen[29] reported for second degree mild burns that the critical energy was 1.09 cal/cm² above the critical temperature of 55°C. Burns which lie between mild second degree burns and severe third degree burns involve injury to the dermis, whose thickness varies from 1 to 4 mm, the average being 2 mm.

To compare burn values based upon Chen's critical energy-critical temperature criteria with deep burn test results obtained from University of Rochester studies, the M.I.T. group calculated the energy needed for total conversion of the collagen in the dermis at 55°C. They considered the fractional wet weight of collagen in the wet dermis to lie between 10 to 30%; they took the heat of fusion in situ collagen to be 24 ± 5 cal/g (i.e. heat for thermal transformation of the collagen involving melting of the crystalline regions). Finally, 55°C was considered the critical temperature for collagen phase transition at the concentration found in the dermal layer.

Working with the University of Rochester data, which included information on the incident radiation intensity at the surface of pigskin and its mean absorptivity, they calculated the rise of the skin temperature for various exposure times and at various skin depths. Plots of temperature rise vs. skin depth were obtained for various intensity levels and exposure times. The areas under these curves were multiplied by the volumetric heat capacity of the skin (0.8 cal/cc °C) to give the enthalpy increase of the skin above 55°C. This critical energy increase (expressed in cal/cm²) ranged from 1.09 for a second degree mild burn to 3.88 for a third degree severe burn. The M.I.T. group then checked this critical energy increase for a third degree severe burn by calculating the calorie uptake which would be required to transform all the collagen in the entire dermis, assuming that the enthalpy increase of the subcutaneous fat is negligible.

The M.I.T. group thus reported on the critical energies required for thermal burns ranging from second degree mild to third degree severe. They provided a physical reasoning for the injury process supported "by a crude qualitative analysis of the value of critical energy required for third degree severe burns". However, they were reluctant to draw any conclusions regarding thermal injuries to complicated skin structure based on this simple treatment, and suggested that further detailed analysis would require a thorough heat transfer analysis of skin. Heat transfer in the dermis should involve a heat-sink at the critical melting temperature to account for heat of melting, in addition to heat absorption due to blood flow. This task was not undertaken in the M.I.T. program because of the complex nature of the skin structure and limitations of time. Instead, considerable effort was directed towards modification of the damage integral criteria originated by Henriques.

5.2.3 M.I.T.'s Modification of Henriques' Burn Injury Criteria

The M.I.T. Fuels Research Laboratory questioned the original format of the Henriques integral criteria for a number of reasons. The Henriques calculation for P, the frequency factor, and ΔE, the activation energy, were made on the basis of relatively low surface temperatures and long exposure times.

The assumption was made by Henriques that the temperature at the basal layer quickly reached its steady state level and was not too different from the surface temperature. However, it was now pointed out that even under steady state conditions, the temperature of the basal layer would be lower than the surface temperature and differences of 1 °C between surface temperature and basal layer temperature would cause a difference of 100% in the damage rate and thus provide an overprediction of Ω (see Fig. 5-2).

In addition, it was necessary to include calculations of the damage during the cooling period, for, as Stoll has shown, damage during the cooling phase, may contribute from 10% to 35% of the total damage in high intensity radiation exposures of skin surface. Also, Henriques calculated the temperature history of the basal layer of the epidermis on the assumption that skin behaves like a semi-infinite solid with constant thermal properties. The nonuniformity of skin structure considering both the epidermis and the dermis, as well as the subcutaneous tissues with isothermal blood core, undermines the assumption of constant thermal properties of the skin. Hence, it was considered that calculation of the damage integral at the depth of 80 microns could not be used realistically to characterize damage at deeper layers of the skin.

Accordingly, the MIT group undertook to modify the Henriques injury criteria as follows: "thermal injury of the epidermis was again assumed to be caused by thermal denaturation of proteins, injury in the dermis was assumed to be caused by thermal shrinkage of collagen, and both processes were assumed to follow first-order kinetics. According to the modified criteria, values of Ω are calculated at various depths in skin and irreverisble damage is said to occur up to the depth where Ω reaches a value of unity. To obtain accurate values of the skin temperature at various depths, an improved heat-transfer model was formulated. Variation of thermal properties in different layers of skin and the effect of blood flow were taken into account. Allowance for the diathermancy of skin for penetrating radiation was made by means of the best avail-

259

able values of absorption coefficients for different layers of skin. A partial differential equation so obtained was solved by a finite difference technique using an IBM 370 computer. The upper limit of the integral in eq. (5.1) was changed to include damage which occurred during the cooling period". Values of P and ΔE were chosen such that predicted values of depth of damage agree with the burn measurements reported by the University of Rochester. The best values of the constants P and ΔE were found to be 1.43 x 10^{72} sec^{-1} and 110,000 cal/mole, respectively for the epidermis and 2.86 x 10^{69} sec^{-1} and 110,000 cal/mole, respectively for the dermis.

Finally, the M.I.T. group undertook a comparison between total energy density transferred to the skin and the depth of damage calculated for a number of GIRCFF fabric burns. Figure 5-9a shows that for thermal dosages to the skin above 4 cal/cm^2 the damage depth is linearly related to dosage. These results are similar to those obtained at Gillette as shown in Fig. 5-9a as is accentuated by plotting the Gillette line on the M.I.T. plot as in Fig. 5-9b. Below 4 cal/cm^2 the depth of damage is constant down to 2.5 cal/cm^2--a feature which may relate to the energy barrier at the dermal-epidermal boundary.

Reference is made to the findings of others[46,49] that as the radiant flux density decreases, the time to produce second degree burns increases, and that to produce a threshold second degree burn, more total energy is required for a low irradiance, and under such circumstances, the skin is heated to a greater depth. Accordingly, the M.I.T. group withholds any generalization of its results, stating simply that "for the heat flux and exposure times involved for the specific fabric burning experiment, the depth of tissue damage is approximately linearly proportional to the total energy input".

5.2.4 Summary

The relationship between the degree of skin burn damage and the level of heat flow to the skin has been studied

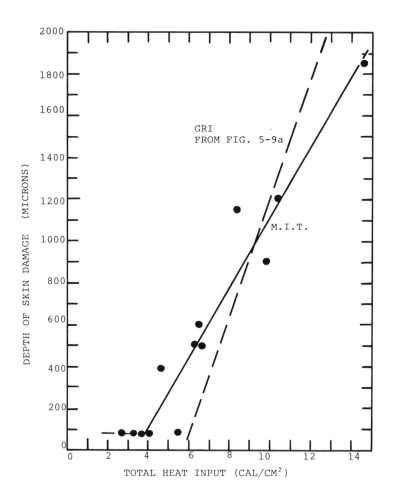

FIGURE 5-9b DEPTH OF SKIN DAMAGE (AS DETERMINED BY M.I.T.)
AS A FUNCTION OF TOTAL HEAT INPUT [R25]

and several criteria for predicting burn damage have been proposed. Henriques et al concluded that local burn injury is a first order rate process (with rate related exponentially to local temperature) and established that during local heating of the skin, its temperature history at the 0.08 mm level can be used to predict the occurrence of a second degree burn, i.e. one involving the full thickness of the epidermis with negligible injury to the dermis.

Noting that tissue damage occurs at all temperatures above 44°C, Stoll modified the Henriques' method to cover the entire exposure period during which the epidermal base (at 0.08 mm depth) was above 44°C, both during heating and cooling. Stoll further demonstrated that temperature rise in skin and the tolerance time to a given damage level depend on the heat (flux) absorbed--and this relationship appears independent of the mode of heat transfer. While refinement of these data was possible to designate first, second, and third degree skin burns due to rectangular heat pulse exposure, Stoll considered it sufficient to designate the damage level of skin areas exposed to flame or high intensity thermal radiation either as severely burned (second degree, or worse), or not burned.

Chouinard, upon analysis of the Stoll data, notes for heat fluxes between 0.3 and 1.0 cal/cm^2 sec (corresponding to that received by the skin from many fast burning fabrics) that a total thermal dosage of 2.0 cal/cm^2 suffices to cause a second degree burn. As a part of the GIRCFF program, Arnold et al at the Gillette Research Institute used the thermal dosage of 2.0 cal/cm^2 to characterize second degree burn areas during garment-on-mannequin burning tests.

In subsequent studies, GRI undertakes a correlation between thermal dosages and depth of skin damage observed in bioptic sections of burned rats and finds that a linear relationship exists between these parameters. The thermal dosage is found to be about 5 cal/cm^2 at a burn depth of 0.10 mm, corresponding to a 2nd degree burn, and about 12 cal/cm^2 for a full dermal thickness injury at a 2 mm depth. On the basis of these findings G.R.I. sets 5 cal/cm^2 as the "criterion for a second degree burn in cases in which the data are not available in a form suitable for the convenient use of the Stoll Criterion".*

The selection of 2 cal/cm^2 or 5 cal/cm^2 as the criterion for occurrence of a second degree burn may seem unimportant in light of the MIT findings that thermal dosages from 2.5 to over 4 cal/cm^2 yield a constant depth of damage at about 100 μm. But if the setting of the 2nd degree burn damage dosage is used to establish the burnt areas of the body and also to calculate the average heat dosage to these areas, then the damage dosage level does become important. For if it is set high, then the average dosage figure in the "burned" area will be automatically weighted on the high side and the calculated % burned area will be on the low side.

The MIT group working under the GIRCFF sponsorship reveiws the Henriques and the Stoll burn criteria and combines their assumptions as a basis for further development of a burn injury model and analysis [R23]. The MIT group also gives limited attention to still another burn criteria, originally proposed by Chen, in which burn potential is expressed in terms of the thermal dosage received by the skin after it reaches 55 °C. This temperature is judged to correspond to the phase transition temperature of collagen at the volume fraction of collagen in a collagen solvent admixture corresponding to that present in the dermis layer.

---

*[R28] page 6.

263

M.I.T. devotes considerable effort towards improving the
Henriques burn criterion and proposes a "modified heat transfer
model for both opaque and diathermanous skin, which allows for
the known variation in properties in depth and the effect of
blood flow (and) is expected to give more realistic and accurate
temperature histories in depth, than those used in previous
studies in which skin was assumed to be a simple, opaque, semi-
infinite solid. The model permits certain modifications, involving the distribution of blood flow in skin and the effect of
changes in blood flow (vaso dilation) due to <u>increase</u> in tissue
temperature, to be made in the future, when such data are available".

For the purposes of the current GIRCFF studies, however,
the M.I.T. group sets aside the effect of local temperature
histories on tissue properties and treats the skin as "an inert
medium, no account being made for any endothermic tissue reactions,
evaporation of moisture and decomposition or carbonization of
tissue. The temperatures thus calculated are higher than would
occur if such processes occurred and the severity of injury predicted from these calculated temperatures are correspondingly
on the high side."

This 'working' criterion is based on the assumption of two
different first-order kinetic processes of thermal injury to
the epidermis and to the dermis respectively. Variation of
thermal properties at different layers are taken into account
along with the effect on blood flow and the presence of the
diathermancy of skin for penetrating radiation, - all to help
formulate an improved heat transfer model for use in calculating
local tissue temperatures at various depths. According to these
modified procedures, values of the Henriques damage integral are
then calculated at various skin depths, and <u>the depth where the</u>
<u>integral reaches unity is designated as the depth of irreversible</u>
<u>damage</u>. The new burn criterion gives good agreement between
bioptically determined experimental measurements and predicted

values of depth of damage based on thermal analysis[R25].

5.3 PREVIOUS STUDIES ON BURN INJURY POTENTIAL OF FABRICS

Much of the work reported thus far on burn injury has had
to do with the exposure of bare skin to various heat sources,
relaying on conductive, convective or radiative heat transfer.
In an effort to provide protections from extreme thermal
exposure, such as an atomic blast, or direct exposure of fire-
fighters to open flame, considerable work has been carried out
on heat transfer from an intense source through textile fabrics
designed as protective materials, i.e. where the fabric is
considered as a thermal resistance. More recently, studies have
been conducted in which the textile fabrics become the heat
source as a result of ignition and the concern is then focused
upon heat transfer from a burning fabric to the wearer's skin.

5.3.1 Fabric Strip Studies

Webster, et al[50] studied the rate of heat transfer from
burning fabrics to a nearby surface with a spacing between
the fabric and the surface of 1.25 cm. In one case, the
receiving surface consisted of a cylindrical iron pipe filled
with water and surrounded by a cylindrical fabric fitted over
a wire cage. In the second experiment, a vertical strip of
fabric was burned at a distance of 1.24 cm from a vertical
asbestos board in which had been mounted a copper block, with
a single thermocouple to measure temperature rise. Webster
studied a variety of fabrics, including cotton poplins, flame-
retardant cotton flannels, viscose fabrics, linen canvas, wool
and cotton blends and wool fabrics of various weights and concluded that "the maximum heat transfer to a nearby surface from
burning fabrics is about 1200 cal/gm, no marked difference
being observed between wool and cotton. The heat dosage,

265

although increasing with the weight of the fabric is, even for
normal lightweight fabrics, considerably above the minimum
value which will cause severe burns. The maximum heat flux
from the flames to the surface is about 0.7 cal/cm^2 sec". In
addition, Webster used the calorific value of cellulose to be
about 4180 cal/gm and that of wool, 4950 cal/gm, as obtained
by earlier studies at the Bureau of Standards, and observed
that the heat "transfer as a fraction of the calorific value
is about one-third for cotton, on the basis of the above
quoted calorific value, and about one-quarter for wool".

Alvares et al[51] measured the heat transfer from
vertically burning cotton fabric adjacent to an aluminum wall.
The specimens were quite large (6 x 1 feet) and the spacing
between the burning fabric and the wall was varied. They
reported that both the total and the radiant heat transfer
reached maximum levels at a fabric-to-wall spacing of 1.5 in,
but the maximum convective heat transfer occurred at a spacing
of 0.75 in. The maximum total heat flux observed at a spacing
of 1.5 in was 0.9 cal/cm^2 sec. Significant quantities of
condensate were recorded at a spacing of 0.5 in and beyond 1.0
in, none were observed.

Brown et al[52] measured the heat transfer vs. time data
when fabrics placed vertically in front of a Transite board
were burned with spacings from 0.25 to 1.0 in. The specimen
sizes here were 25 by 22 in, and the results showed that
heat transfer at a spacing of 0.5 in was most hazardous in
terms of the length of time to occurrence of second degree burn
equivalence. It was found that all fabrics tested which burned
uniformly provided heat transfer at a level about double that
required to cause second degree burns.

Robertson[53] studied heat transfer from burning fabrics
and films employing spacings of 5/16 and 1.0 in and for
orientations from 0-30°. His results indicated that at the

closest spacings a major portion of the thermal transfer takes place through convection and vapor condensation, while at the higher spacings, radiation plays an important role. He encountered difficulties in studying thermoplastic materials because of melting, dripping, and irregularity of burning.

Generally, it was observed that where test fabrics burned irregularly, the total heat transferred to nearby surfaces was less than for fabrics of equivalent weight which burn uniformly and completely. But it remained to evaluate heat transfer in local burning areas, particularly when such burning was accompanied by melting and local dripping.

Recently, Chouinard[34] tested the hypothesis that thermoplastic fibers may transfer substantially more heat than non-thermoplastics in the local areas where they burn aided by conductive transfer from dripping molten polymer. His experiments involved the burning of 45° inclined specimens spaced 1/4 in from a parallel sensor board containing numerous sensor disks. Nylon and polyester fabric samples were parallel stitched with cotton thread to enhance flame spread as a cotton sewn seam might do in actual garments. Alternatively, a 1 oz/yd² cotton scrim support was also used to enhance flame propagation and was so porous as to permit melted polymer to fall down to the sensor board. The heat of burning of the cotton was subtracted from the energy density readings of the experiment. The net energy density reading (described in Section 5.2.2) was normalized with reference to fabric weight (see Table 5-3) yielding the Base Burn Potential. The data indicate that cotton transfers the highest energy density per unit weight while polyester produced the smallest transfer _despite_ the polymer melting and dripping. The others lie between, and it was noted that "fragment dropping acrylic, nylon, or polyester fabrics transferred smaller percentages of their heats of burning, H_b, than cotton or PE/C fabrics, which remain in place as they burned".

267

Table 5-3

<u>HEAT TRANSFER TO SENSING BOARD</u>

Fiber	Base Burn Potential $(cal/cm^2 \ oz/yd^2)$	Support
Cotton	1.7	---------
Nylon	1.4	Stitched
Nylon	1.5	Scrim
Wool	1.3	Scrim
PE/C	1.3	---------
Acrylic	1.2	---------
Polyester	0.7	Stitched
Polyester	0.8	Scrim

It was further noted that "heat transfer from double layers of cotton or PE/C fabrics was approximately the sum of transfers from individual fabrics. When cotton and thermoplastic pairs were tested, the order of layers had a marked effect on transfer. ......When the thermoplastic fabrics were underneath, heat from the relatively fast burning cotton caused polymer droplets to form, which fell to the sensor surface and did not burn completely. The heat generation potential of the thermoplastics was not fully realized, and the average transfer increased only slightly because of their presence". Similar behavior was observed in tests with cotton dresses worn over nylon slips on the mannequin as reported below.

When the thermoplastic was placed above the cotton fabric during the test, the thermoplastic fabric first melted onto the burning (but supportive) cotton and then burned in place. For such sensor board tests the sum of the heat transferred from single layers of cellulosic and thermoplastic very nearly equalled that transferred from the double layer (with the thermoplastic on top). Burning tests with thermoplastic dresses worn over cotton slips were not reported by Chouinard.

5.3.2 Mannequin Studies

Chouinard et al[34] used a heat sensor instrumented mannequin to evaluate heat transfer from burning garments to various parts of a fixed simulated body. His experimental fabrics included cotton, acrylic, polyester, and nylon fabrics, and a polyester cotton blend. The heat of combustion H_c, and the heat of burning, H_b, of such materials have been measured by Yeh[18]. These are reported in Table 5-4.

Chouinard reported that cotton fabrics liberate nearly constant energy per unit weight over a wide range of fabric weights and pointed out on the basis of Table 5-4 that cotton

269

Table 5-4

HEAT GENERATION CAPACITY OF TEXTILES [18]

Fabric	H_b cal/g	Rate cal/sec-cm	H_c cal/g	H_e H_b/H_c %
Cotton	3315 $\neq$	78.3	3688	89.9
Acrylic	4118	43.7	7020	58.7
Polyester*	2188	17.5	5255	41.6
Nylon*	3715	13.2	6926	53.6
Polyester/Cotton	2776	69.5	4907	56.6

H_b	heat of burning in air at 21% oxygen
Rate	of heat generation per unit sample width
H_c	heat of combustion in oxygen bomb calorimeter
H_e	H_b/H_c, or percent of H_c liberated by burning in air
*	cotton gauze underlay required as ignition source
$\neq$	the experimental significance of these figures is not known.

and cotton/polyester had relatively high rates of heat generation and nylon and polyester fabrics the least (actually requiring a cotton source to sustain ignition). But it should be noted that Chouinard "rate" values are based on the heat generated per unit width of the fabric, not per unit weight. The H_b and H_c columns are based on fabric weight and their ratio is shown in percentage in the H_e column. Here it is shown that the all cotton liberates 90% of of its theoretical heat, the PE/C blend 57% and the Polyester 42%. For a 3 oz/yd^2 fabric, Chouinard points out that the heat generation capability will vary from 22 cal/cm^2 for a PE fabric to 41 cal/cm^2 for an acrylic fabric. And since the threshold energy absorption for second-degree injury is approximately 2 cal/cm^2 at the heat flux levels found in burning fabrics - he concludes that all these materials can cause second-degree burns if a significant portion of the heat of burning, H_b, is absorbed by the skin.

Chouinard's garment tests were used first to study heat transfer properties of single dresses (with no underwear) of cotton, polyester/cotton and acrylic. Record was kept of the time, t_1, from ignition of the dress until the 2 cal/cm^2 epidermal tolerance was exceeded at any one sensor on the mannequin. Since lighter weight garments burned faster, smaller values of t_1 were observed for such materials.

Chouinard measured the total energy transferred to the sensor points at which the tolerance limit was exceeded during the entire burn. He then normalized this value for each fabric to the total energy transferred by a standard 3.2 oz/yd^2 cotton fabric. The resulting ratio was named the Heat Transfer Index. The 3.2 oz. cotton dress was observed to transfer an energy density of 8.0 cal/cm^2 over 14 sensors on the mannequin body.

Cotton, polyester/cotton and acrylic fabrics were found to lie in the same Heat Transfer Index (HTI) grouping, but wool was substantially lower in HTI, while nylon and polyester single

271

dresses self-extinguished because of melting and dripping before triggering any sensors, thus leading to the zero HTI rating. As Chouinard pointed out, these fabrics "appear unlikely to produce trauma over more than quite a limited area unless used in com- bination with other types of materials which can overcome their inherently low flame-transmission properties". It should be noted that in Chouinard's tests the final area of the garment burned and the energy per unit area transferred increased with fabric weight for each material. Thus, the HTI is dependent on the area extent of flame propagation in the unique geometric interaction between a particular garment and the dummy, and it is dependent as well on the mass per unit area of the material. But it does give an arbitrary measure for an immobile wearer, of the extent of burn suffered compared to the case of the burning 3.2 oz. cotton dress.

Similar tests were run with combinations of dresses with underwear--i.e. using 4.3 oz cotton and 2.2 oz nylon slips worn under 3.2 oz and 7.0 oz/yd² cotton dress materials. It was observed that the "burn area"- time relationship (a parameter similar to flame propagation velocity) reflected that the rate of burn area spread was only slightly less for the cotton dresses with underwear vs. for the cotton dresses alone. Adding cotton underwear to the cotton dresses resulted in a larger total area of burn involvement (at the end of the test) and also a higher energy density at these points. Addition of nylon underwear to cotton dresses showed little effect on these test parameters, but after such tests unburned droplets of nylon were observed on the lower half of the mannequin. Generally it was noted that several layers of cotton garments combined transferred an energy density equivalent to that of a single cotton dress of the same total weight as the combination.

5.4 <u>BURN INJURY POTENTIAL OF THE GIRCFF FABRICS IN GARMENT FORM</u>

It was considered desirable to focus on the Chouinard results as a prelude to discussion in experimental work of the Gillette Research Institute since in much of the G.R.I. work reported here the methodology was based on the Chouinard experience, and Du Pont data collection equipment was used for the experiments. It will be recalled that the G.R.I. group studied the <u>ignition</u> potential of the GIRCFF fabrics in simulations of typical kitchen stove "sleeve" accidents. Their results were described in Chapter III. This section reports on the G.R.I. simulation of real life accidents involving burning clothing constructed from GIRCFF fabrics. In these studies prototype garments were burned on mannequins instrumented with calibrated heat flux sensors and, in some cases, with temperature sensitive papers. The results reported included the total heat input at sensor locations, the relative torso area which would have suffered second-degree (or worse) burns, and the calculated depth of burn which would have been obtained from the heat inputs recorded.

5.4.1 <u>Overall Heat Transfer from Burning Single Garments</u>

The GRI experiments focused on the GIRCFF fabrics in single garments and in various combinations. In addition, a few fire resistant fabrics were included toward the end of the program. The results of experiments in which mannequins were clothed with simple A-style dresses (no slip) and fires initiated by ignition at the front of the dress indicated that:

1. "Cotton and PE/cotton (65/35 or 50/50) dresses burn extensively and give out enough heat to cause second degree and deeper burns (heat inputs to skin greater than about 2.5 cal/cm^2) over 70% or more of the torso and head, if allowed to burn to completion."

273

2. "Nylon, polyester, NOMEX and FR cotton fabric dresses do not sustain combustion by themselves even when ignition is attempted with a large paper fuse."

3. "For fabrics of comparable weight, the ultimate areas of burn are similar for cotton and PE/cotton blends, but the cotton burns more quickly; for a given fabric type, the duration of burn increases linearly with fabric weight."

4. "The average heats transferred to burned areas are similar for cotton and PE/cotton blends of similar weight and they increase linearly with the fabric weight."

5. "The general pattern of flame spread following the ignition of cotton or PE/cotton blend dresses at the front of the hem, is for the flames to spread up the front from close to the point of ignition, but not always affecting the front of the shoulder, then to spread relatively slowly to the back, usually cutting off large portions of unburned fabric at the bottom which fall away, eventually leaving only a relatively few of the instrumented areas unaffected (i.e. heat inputs of less than 2.5 cal/cm^2)[R28]."

Since the G.R.I. results have been reviewed, in part, in Chapter IV on "Flame Propagation", only the G.R.I. conclusions dealing with heat transfer and burn damage will be discussed here. The fifth conclusion quoted above which concerns the pattern of flame spread has been previously discussed in Chapter IV and illustrative data are shown in Figs. 4-10 to 4-12.

In interpreting the experimental results, G.R.I. relied on the Stoll criterion for second degree burns as suggested by Chouinard, i.e. they considered any density of energy transfer from fabric to skin above 2.0 cal/cm^2 to be sufficient to cause

a second degree burn. As stated earlier, when data were not
available in a form suitable for convenient use of the Stoll
criterion, G.R.I. relied on their own pathological determinations
of epidermal and dermal necrosis as shown in Fig. 5-9a. On
this basis energy transfer of 5 cal/cm^2 corresponded to a
second degree burn (i.e. a burn depth of 100 µm) and
12 cal/cm^2 to a third degree burn (i.e. a depth of 2000 µm).

In addition to using heat flux readings to determine
areas of second degree burns (and higher) during garment fires,
G.R.I. also calculated the average heat transferred to burned
areas. Data shown in Fig. 5-10 provide the basis for the
conclusion that the average heats transferred by cotton and by
polyester/cotton fabrics are similar and increase linearly
with fabric weight. The data reported by Chouinard[34] for
cotton, PE/C, acrylic, and wool fabrics have been also plotted
in Fig. 5-10. Chouinard's data were average readings of 13 to
17 sensors which had reached 2 cal/cm^2 energy density or above,
with the exception of the wool data, which were based on a few
sensors only. The wool dress burned only to the point where
one or two sensors recorded second degree injury, whereas the
100% cotton, PE/cotton and acrylic dresses burned on to cause
second degree burns over 50 to 70% of the torso sensor
locations.

The two linear relationships shown in Fig. 5-10--one for
the data obtained by G.R.I. and one for the Chouinard data--
are separated by an energy density of about 5 cal/cm^2, a
magnitude of itself sufficient to cause second degree burns.
A possible reason for the difference in the results obtained in
the two sets of garment burning experiments under the same test
conditions has been discussed in Section 5.2.4.

Other experimental differences will affect the results as
well, for example use of different sensors, different mannequin
sensor locations and different garment designs. In short, Fig. 5-10
is in no sense absolute properties of the fabric tested. However,
taken separately, both the G.R.I. data and the Chouinard data indicate

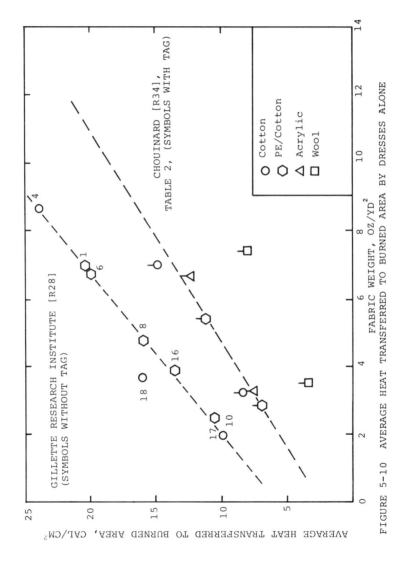

FIGURE 5-10 AVERAGE HEAT TRANSFERRED TO BURNED AREA BY DRESSES ALONE

that heat transferred in a garment burn test of dresses made
from cotton fabrics and of polyester/cotton blends is a
linear function of fabric weight, and not dependent on the
fiber composition. Furthermore, even for the lightest fabrics
tested, the level of heat transferred is sufficient to cause
second-degree burns or more over 70% of the mannequin torso
area.

GRI considered these results to be unexpected since
significant differences in the heats of burning of cotton and
of the PE/C blends had been observed by Yeh [18] in calorimetric
studies. Yeh's data recalculated by GRI in terms of heat
released per unit area (i.e. cal/cm^2) are included in Table 5-5.
The last two columns of Table 5-5 show estimates of the heat
transferred per unit area in the GRI garment tests in cal/cm^2
and as a percentage of the heat released as reported by Yeh.
As GRI points out, the percentages of heat transferred per unit
area lie between 25% and 50%, the percentage increasing as the
weight of the fabric decreases. (Note that this is a comparison of garment burn heat transfer versus heat of burning
in air, not heat of burning in an oxygen bomb.)

It should be noted that Yeh's calorimetric data reflects
a relatively constant burning heat per unit weight for cotton
fabrics of varying weight. However, Yeh interpreted the
results as showing some effect of fabric weight and structure
for polyester/cotton fabrics. In the case of garment burning,
different levels of relative heat transfer are observed even
for 100% cotton fabrics with higher levels of relative transfer
in lighter fabrics.

Finally, it is useful to compare the heat transfer in a
garment burning test with the heat of combustion in an oxygen
bomb as reported by Chouinard on the basis of Yeh's results
(see Table 5-4). By multiplying the percentage H_e values
given in Table 5-4 with the percentage heat transferred shown

277

Table 5-5

HEATS REALIZED BY BURNING GIRCFF
FABRICS UNDER NORMAL ATMOSPHERIC CONDITIONS
(BASED ON DATA FROM YEH, REFERENCE 18)

Fabric Number	Fiber and Finish	Fabric Weight (oz/yd^2)	Heat Released Per: Unit Weight (cal/g)	Unit Area* (cal/cm^2)	Heat** Transferred in Garment Test	Percent
4	Cotton	8.7	3200	94	23	24
1	PE/cotton,65/35,DP	7.1	2600	64	20	31
6	PE/cotton,65/35	7.0	2790	67	20	30
15	PE/rayon, 65/35,DP	6.6	2400	54	--	--
8	PE/cotton,65/35	4.8	3160	51	16	31.5
5	Cotton	4.0	3160	43	--	--
16	PE/cotton, 50/50	3.9	2780	37	13	35
18	Cotton	3.7	3270	41	16	39
17	PE/cotton, 65/35	2.5	2970	25	11	43
10	Cotton	1.9	3120	20	10	50

*Calculated from Yeh's data assuming complete combustion.

**Estimated from Figure 5-10.

in Table 5-5, one infers that heat transferred in garment tests
will range from 22 to 45% of the combustion heats (in oxygen)
for cotton fabrics, and from 17 to 25% of the combustion heats
for polyester/cotton fabrics. The calculation of such heat
transfer efficiencies is useful only in comparing structural
designs of materials made from a given fiber since the heats
of combustion in oxygen vary from fiber to fiber (as seen in
Table 5-4). However, in the final analysis, the important
parameter is the absolute energy transferred, since injury
depends on the actual heat flux reaching the skin.

5.4.2 Local Heat Transfer from Burning Single Garments

In addition to reporting average energies of heat transfer,
the GRI group prepared curves of heat input distribution for
cotton dresses burned on the mannequin. These curves indicate
the area of the torso receiving various heat inputs at 15, 30,
60, 120 and 180 seconds after ignition and at the end of the
burn (if different from the last time-designated curve).
Typical curves are shown in Fig. 5-11a and 5-11b for fabrics
of widely different weights. For the lighter weight cotton
fabric #10 (1.9 oz/yd^2) heat input is more rapid over a larger
torso area than for the heavier cotton fabric #4 (8.7 oz/yd^2).
The difference in the level of heat transferred is obvious,
with virtually no torso area being subject to more than 20 cal/cm^2
in the case of the 1.9 oz/yd^2 fabric test, while a considerable
area of the torso receives more than 20 cal/cm^2 in the test
of the 8.7 oz/yd^2 fabric.

As pointed out earlier, the GRI group considered that transfer exceeding 5 cal/cm^2 would cause epidermal necrosis to the
depth of 100 μm and heat inputs in excess of 12 cal/cm^2 would
cause damage to a depth of 2 mm. Thus, Figure 5-11 is useful in
that it shows that average heat transferred during a burning
test, as shown in Fig. 5-10 reflects only part of the story
since heat inputs to local areas may far exceed the average.

279

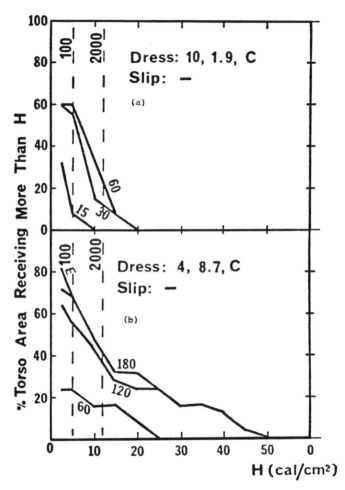

FIGURE 5-11 HEAT INPUT DISTRIBUTION CURVES FOR COTTON DRESSES
BURNED ON THE 6X MANNEQUIN. CURVES ARE FOR 15 AND 30 SECS
AFTER IGNITION, IF DETECTABLE; AND FOR 60, 120 AND 180 SECS,
IF DIFFERENT FROM CURVE FOR PREVIOUS TIME. THEY ARE RESULTS
OF SINGLE BURNS. THE BROKEN LINES REPRESENT HEAT INPUTS
BELIEVED TO CORRESPOND TO BURN DEPTHS OF 100 AND 2000 μM.
THE GARMENT FABRIC NUMBERS, WEIGHTS (OZ/YD²) AND FIBER
CONTENTS ARE INDICATED [28]

Still another representation of rate and extent of burn injury is presented in Fig. 5-12. Here, the heat inputs recorded by various sensors on the mannequin are shown for arbitrarily selected times during the course of the burn. G.R.I.[R28] suggests that "while it cannot be expected that exactly similar heat inputs would occur from a real garment burning on a live child, the information is useful in providing an indication of the pattern of burn injury and the potential depth of that injury which might be suffered by a child wearing a dress which was ignited at the front". The main features of these observations were outlined by G.R.I. as follows:

o the lateral flame spread around the side leading
   to eventual burning of the back of the dress,

o the tendency is for highest ultimate heat dosage
   to be localized in the front and below the waist
   (though Fig. 5-12 shows that high inputs can also
   occur on the back),

o the usually lower heat dosage on the back is
   associated with large pieces of fabric dropping
   away before burning and the lower heat inputs on
   the upper part of the body may be caused by
   mannequin quenching of the fabric,

o the very short time needed for the lightest fabric,
   the 1.9 oz/yd^2 cotton (No. 10), to liberate most
   of its heat, as compared with the relatively long
   time needed for the heaviest fabric, the
   8.7 oz/yd^2 cotton (No. 4), is a manifestation of
   the different rates of flame spread which have been
   discussed in Chapter IV,

o the initial flame spread up the front from the point
   of ignition is accompanied by a substantial heat
   input to the throat and presents a real danger of
   inhalation of hot gases in actual life accidents,

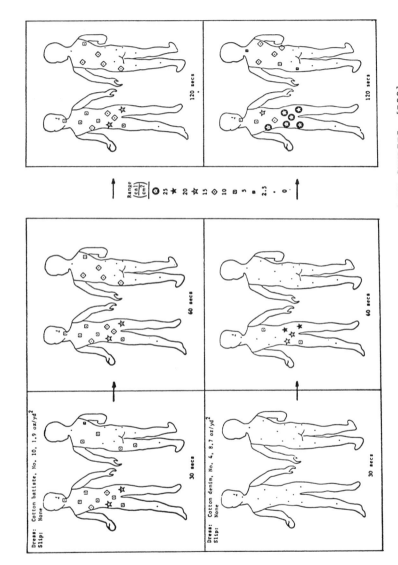

FIGURE 5-12 RATE AND EXTENT OF BURN INJURY IN MANNEQUIN STUDIES [R28]

5.4.3 Fabric Combinations and Garment Designs

In addition to studying the average heat transferred from burning single dresses, the G.R.I. group also determined the average heat transferred to burned areas from various dress/slip combinations of differing composition and total weight. These experiments included fire resistant cotton and NOMEX fabrics. The results of these experiments are first summarized, then discussed in detail in this section.

The conclusions reported by G.R.I. are quoted below:

o "Dress/slip combinations act similarly (in terms of area burned, burn duration, and average heat transferred to the burned area) to an all-cotton or a PE/cotton blend dress of a weight which equals the combined weights of the dress and slip fabrics, provided the fabrics are themselves cotton or PE/cotton. The data for dresses of various weights of (a) cotton and (b) polyester/ cotton thus provide useful references with which to compare the data for dress-slip combinations".

o "If a garment (dress or slip) of a flame retardant fabric is burned in combination with a garment of flammable fabric (e.g. PE/cotton), the results depend on which fabric is on the inside and which outside. If the FR garment is on the inside, the average heat transferred to the burned area and the area of burn are less than would be the case if it were not present; if the FR garment is on the outside, it appears that the average heat transferred is greater than that for the inner garment alone, the burn duration is slightly increased, and the area of second degree burn is slightly reduced."

o "The use of a belt provides some protection by
slowing down the flame spread, increasing the
duration of burn and reducing the area of burn."

o "A 2.5 oz/yd^2 nylon slip under a cotton or PE/c
increases the average heat transferred to the
burned area by an amount which appears to increase
significantly with the weight of the dress fabric.
The nylon slip increases the duration of the burn
over that for the dress alone."

o "A polyester dress burned with a PE/c blend slip
appears to give similar results to a NOMEX or
FR cotton dress over a PE/c blend slip, both in
terms of average heat transferred and area of
burn."

o "Panties provide no significant protection against
burn injury, though a bra may provide some."

The data plotted in Fig. 5-13 show that heat transfer from
burning a garment combination is very nearly the same as that
from burning a single garment of weight equivalent to the
combination, provided that both fabrics burn freely when tested
alone. The heat transfer vs. fabric weight relationship
obtained in tests on single dresses is shown in Fig. 5-13 as
a dashed line, and it is seen that cotton/cotton combination
10/10 nearly falls on this line, as does the cotton/PE-c
combination 4/17. However, when a thermoplastic is used
as the slip material (e.g. #12, nylon tricot) together with a
PET/cotton dress, the heat transfer at a given weight is less
than for the all-cotton or cotton/PE-c combinations. This
difference becomes less at higher dress weights and, indeed,
the cellulosic/nylon combinations lie on a straight line which
intersects the single garment dashed line at about 11 oz/yd^2
combined weight. This suggests that when more cellulosic
fuel is provided, the nylon slip will burn more completely.

284

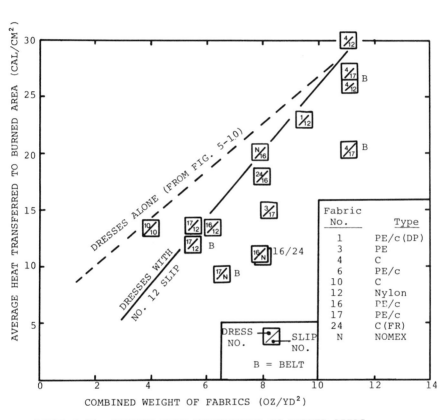

FIGURE 5-13   AVERAGE HEAT TRANSFERRED TO BURNED AREAS
BY DRESS-SLIP COMBINATIONS   [R28]

Figure 5-13 also contains interesting data regarding the behavior of fire resistant materials when used in combination with more flammable materials. The particular combinations to note are those containing NOMEX fabric #N (4.1 oz/yd^2), the fire resistant cotton #24 (4.0 oz/yd^2), combined with the polyester/cotton #16 (3.9 oz/yd^2). Figure 5-13 reveals that when a dress of NOMEX (N) is worn outside and a slip of PE/c #16 inside, the average heat transferred is almost double that when NOMEX is used as the slip material and worn inside. And when a dress of fire resistant cotton #24 is worn outside with a slip of PE/c #16, the heat transferred is about 1.70 times that transferred when the fire resistant cotton is used as the slip material and the PE/c is the outer garment. Also, one notes that for these two pairs the arrangement corresponding to higher heat transfer gives results that fall on the solid line for the combinations of all cellulosic or of PE/c blend dress fabrics (e.g. #17, PE/c; #1, PE/c, and #4, C) with a thermoplastic slip fabric (e.g. #12 nylon tricot), see Fig. 5-13.

A final observation recorded in Fig. 5-13 is that the addition of a belt to the dress design reduces average heat transferred to a significant degree. However, while the effect of the belt is to limit the total area burned and to reduce the average heat transferred, its presence tends to slow down the speed of flame propagation and, in many cases, to increase the duration of the burn.

The discussion of the G.R.I. test results of fire resistant fabrics worn in conjunction with flammable fabrics in dress-slip combinations gives some hope of alleviating burn damage when the fire resistant material is worn as the inner garment. But, in the final analysis, the data in Fig. 5-13 reveals that even with this improved behavior, the average heat transfer of the garment combinations in question is still far in excess of that which causes second-degree burns.

The full benefit of fire resistance in a given fabric is realized only when that fabric is used in a single layer garment, i.e., worn alone. Figure 4-12 in the preceding chapter has shown that zero area of second-degree burn is observed in tests of single dresses made from fire resistant cotton (#24) or Nomex (R). The nylon tricot (#12) and the polyester double knit (#3) likewise failed to sustain combustion when worn alone, and therefore, recorded no heat transfer. In contrast, all cotton and PE/C which propagated flame, transferred enough heat to cause second-degree burns or worse.

In the practical circumstances of garment fires, the transfer of heat from burning fabric to skin takes place only if the flame propagates over the garment. If the flame self-extinguishes either at the very beginning of the test, or at a later stage when "critical" fabric-skin spacings serve to extinguish the flame, the total heat transferred will be considerably reduced, and the heat transferred per unit area may also be reduced. Thus, the heat transfer data recorded by GRI inherently includes the effects of flame propagation, and of all the variables which influence flame propagation in a garment test.

5.5 BURN INJURY POTENTIAL OF GIRCFF FABRICS IN STRIP FORM

The Fuels Research Laboratory of M.I.T. undertook to measure the heat transferred from burning GIRCFF fabric strips to a skin simulant. The observations made in these experiments regarding the nature of flame propagation at various orientations and fabric-simulant spacings have been discussed in Chapter IV. A brief review of the observed burning modes will show some aspects of burning behavior which may not have an effect on speed or extent of flame propagation

287

but which will generally have a significant effect on heat transfer. An example is the dripping of polymer melt onto the skin simulant in burn tests of 100% thermoplastics and also of some PE/c blends: In some cases, the melt would continue to burn after falling on the simulant and still, in others, the burning melt would separate from the main fabric, and self-extinguish. In some cases, notably the heavier double knit polyester and the wool sample, a very narrow flame propagated, leaving most of the specimen unburnt. Because of these numerous variables, there could be no certainty of accuracy and reproducibility of heat transfer data obtained in tests on thermoplastic or on wool samples. The M.I.T. group, therefore, worked primarily with the cellulosic and cellulosic blends.

In preliminary tests, the skin simulant was placed beneath the fabric and allowed to burn without control of initial temperature, humidity and wind draft. The heat transfer data obtained were used to calculate the Henriques damage integral without correcting for initial skin temperature. In later tests the temperature measurements in the simulant were normalized to 37°C and the temperature history at various skin depths was calculated as described earlier in this chapter. These latter data provided the basis for using the modified damage integral in calculating the depth of damage which would be suffered by human skin exposed to the observed heat flux.

5.5.1 Skin Simulants

The use of human skin is obviously not feasible in garment burn injury studies. Furthermore, the use of animal skin for correlation of depth of burn damage with heat dosage or time temperature history is a prolonged and expensive technique. Therefore, it is highly desirable to use a skin simulant in the proximity of burning fabrics--both from the viewpoint of flame propagation and also from consideration of simulated burn injury.

"The guiding principle in the search for a skin simulant is the necessity that the simulant accept heat at a rate identical to the acceptance rate in skin itself when the surfaces of the two are subjected to identical histories of temperature or of heat flux density. With this condition met, the (burning) cloth behaves identically over skin or its simulant, in all respects: heating, vaporization, and condensation of moisture and pyrolysis products, melting and solidifying, etc."[R25]

The necessary and sufficient condition for simulation is that the thermal inertia of real skin and of the skin simulant be the same[29], i.e.

$$(kc\rho)_{simulant} = (kc\rho)_{skin} \qquad\qquad (5.2)$$

where k is the thermal conductivity, c is the specific heat and ρ the density.

Transite has been found to be a good skin simulant and has been used in many parts of this program as the body placed in the proximity of burning fabrics.

The insertion of sensors and thermocouples in the surface of the Transite has made it possible to measure the heat flux impinging on the simulant surface. However, measurement of the temperature at 80 µm in depth, the zone of major importance in burn injury, was virtually impossible because of the dimensions of the thermocouple bead. Thus, it was found necessary to use a skin simulant which "stretched" its normal dimension, here denoted by X.

This was done by selecting appropriate values of thermal conductivity for simulant while still satisfying the requirement of equation (5.2). Thus the stretch factor, a, was

$$a = \frac{X_{simulant}}{X_{skin}} = \frac{k_{simulant}}{k_{skin}} \qquad (5.3)$$

Since there appeared to be no single homogeneous material to satisfy equation (5.2) together with a high value of stretch factor, it was found necessary to use a combination of copper and air. Such a design was reported by Chen [29] for a study of heat and moisture transfer to skin. The design consisted of an arrangement of parallel copper air-spaced fins attached to a surface plate. The MIT group redesigned this system for the current study using a rod-type copper air skin simulant. And, with the given stretch factor, they were able to insert thermocouples at an equivalent of 80μm for monitoring time-temperature histories at this level for application to the Henriques integral. Other experiments with a variety of skin simulants were reported [R25].

5.5.2 Effects of Fabric - Skin Spacing and Gravitational Orientation

The MIT group examined the influence of fabric-simulant spacings on heat flux for fabric burning in horizontal orientation, at 45° and at 90°. In horizontal burning, they observed that for three fabrics tested, the thermal dosage to the skin reached a maximum at 1/2" spacing. At a spacing of 1/8" all three fabrics failed to support combustion presumably for lack of sufficient air. At a spacing 1/4" the 3.7 oz/yd^2 cotton fabric (#18) did support combustion but the others did not, i.e., #17, the 2.53 oz/yd^2

PE/C, and #6 the 6.99 oz/yd^2 PE/C fabric. All three fabrics sustained combustion at 3/8" spacing. It was reasoned that "fabric-skin spacing represents one of the most important parameters from the injury point of view. Spacing between garments and body changes from almost zero to more than an inch depending on the type of garment and area of the body. During a garment fire, the spacing usually changes because of factors like chimney effect, curling, softening, sticking of the fabric, etc. and results in a range of injuries at different parts of the body."

The study of the effect of spacing included experiments in 90° upward burning mode -- the most damaging one. The results have been shown as a part of the discussion of flame propagation in Chapter IV -- (see Fig. 4-8 for the 2.0 oz. cotton batiste, (#10) and in Fig. 4-9 for the 3.7 oz. cotton flannel, (#18)). It was pointed out that flame speeds first increase monotonically with spacing, then levels off. However, skin damage is greatest at the minimum spacing of 1/8", decreases as the spacing is increased, and reaches a minimum of about 9/32". It subsequently rises and reaches a maximum at a spacing of about 15/32", then decreases as the spacing is further increased.

The MIT group evaluated the effect of spacing on heat transfer as follows: "The factors affecting the total thermal dosage and injury to skin are: (1) radiation from hot gases and fabric; this is directly proportional to the fourth power of the absolute temperature and inversely proportional to the square of spacing; (2) convection and conduction through the gas phase; this is directly proportional to the temperature of the hot gases above that of skin, and is a direct function of the gaseous recirculation; (3) condensation of steam and pyrolysis products; this becomes

291

significant at a certain minimum spacing and increases with
a decrease in spacing until a critical spacing below which
flame spreading cannot be sustained; (4) exposure time, which
is inversely proportional to flame speed and thus decreases
with increase in spacing.

"The effect of spacing presented in Figs. 4-8 and 4-9
can be explained on the basis of heat transfer considerations.
Below a spacing of 9/32 in, not enough air is available for
complete combustion and the temperatures of the hot gases and
fabric are lower, but flame quenching starts at these spacings,
resulting in condensation of steam and pyrolysis products; heat
transfer due to condensation keeps on increasing until a
spacing lower than 1/8 in is reached; this factor is large
enough to make the damage at 1/8 in higher than that at all
other spacings studied. Above a spacing of 9/32 in, condensation
is negligible, but due to increase in air circulation, the
efficiency of combustion and hence the temperature, increases
and these two factors combined are sufficient to increase the
thermal punishment until a maximum at about 15/32 in is reached.
Beyond this spacing the temperature of the gases and air
circulation tend to level out, and the increase of heat
transfer path, combined with decrease of exposure time
(because of increasing flame speed), tend to reduce the damage
with further increase of spacing."

A summary of the burn tests at 0°, 45° and 90° for 1/2 in.
spacing has been shown in Table 4-2 for 100% cotton and for
PE/cotton blends. Flame speeds at 45° and 90° have been
reported to be about an order of magnitude higher than for
horizontal burns. Thermal injury caused by most fabrics
(as shown in Table 5-6) also increases as the orientation
goes from 0° to 45° to 90°. Only fabric #5, a 100% cotton

Table 5-6

EFFECT OF FABRIC-SKIN ORIENTATION ON THERMAL INJURY; 1/2 INCH SPACING [R25]

Fabric Number	Weight (oz/yd^2)	Depth of Damage in Skin (microns)		
		Horizontal	45° Up	90° Up
Cotton				
10	2.0	80	300	390
18	3.7	500	800	1280
5	4.0	650	1600	1380
9	7.4	1650	2800	3000
4	8.7	2100	3200	3350
PE/Cotton				
17	2.5	80	80	80
16	3.9	80	560	850
8	4.8	550	680	1000
15	6.6	1900	900	1150
6	7.0	600	900	2150
1	7.0	2800	1700	2200

and PE/c fabrics #1 and #15 behaved differently. In the
case of #15 and #1 melt-dripping was assumed by the M.I.T.
group to cause unusually high heat transfer in horizontal
burning.

5.5.3 Horizontal Strips

Skin damage depth resulting from horizontal burning
experiments with a 1/2 in spacing is shown in Fig. 5-14 where
it is plotted against fabric weight. Clearly, the heavier the
fabric burned, the deeper the skin damage, but the cotton
fabrics cause deeper damage than do PE/c fabrics of equivalent
weight. It was pointed out previously in Chapter IV (Fig. 4-5)
that the PE/c fabrics burned faster than the 100% cotton fabrics
of equivalent weight, and this explains the lower depth of damage
caused by the PE/c fabrics. The fact that PE/c #16 and PE/c #17
gave equivalent depth of damage at 80 μm, even though one was
60% heavier than the other, was explained by assuming that the
difference in thermal dosage given by these two fabrics was
actually smaller than the stepwise increase in thermal dosage
"required for the damage to propagate from the epidermis into
the thermally less sensitive structure of the dermis.....".

The greater burn damage from the 100% cotton fabrics as
compared to that from PE/c fabrics of the same weight could
also be explained by reference to the values for calorimetric
measurements of heat of burning (in air), which were found to
be (see Table 5-4) 3320 cal/g for 100% cotton and 2780 cal/g
for PE/c or, in other words, a higher heat of burning in air
for all-cotton fabrics.

It is worth noting and perhaps significant that fabric
#1, the durable press treated 7.07 oz PE/c twill slack material,
transferred significantly more heat than did the

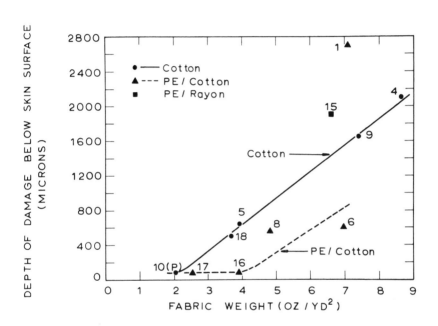

FIGURE 5-14 EFFECT OF FABRIC WEIGHT ON
BURN INJURY, HORIZONTAL BURNING,
1/2-INCH SPACING [R25]

295

equivalent non-treated fabric #6 with a burn depth of about 2700 μm for the former and 600 μm for the latter.

The other durable press treated fabric (#15 PE/rayon) is also seen in Fig. 5-14 to transfer considerably more heat than the untreated PE/cotton of corresponding weight (#6). Yet, fabric #15 was reported by Yeh to give off 2400 cal/g in calorimetric burning, while #6 gives off 2790 cal/g.

If one traces the behavior of DP treated fabric #1 (PE/c) and of untreated #6 in GRI's garment tests (see Fig. 5-10), the relatively higher heat transfer from #1 as compared to #6 is not observed. Thirty seconds after ignition, the area burned to the second-degree level is nearly the same for the two fabrics; after 60, then 90, seconds, the second-degree burn area of fabric #6 is even somewhat greater than for fabric #1, and the total duration of burn for dresses of fabrics #1 and #6 are virtually the same. Thus, the higher heat transfer values for the DP treated PE/c and PE/rayon blends in fabric strip burning tests at MIT cannot be verified in the garment burn tests at the GRI. In view of this discrepancy, it is instructive to look back at Chouinard's results, for he tested the same fabrics in panel form and in garment burn tests. The results plotted in Fig. 5-10 for heat transfer in garment burning for all cotton, PE/c, and acrylic fabrics have shown that all three materials fall on the same line relating average heat transferred to fabric weight, indicating that fiber difference is not important here. Yet, when Chouinard burned a range of fabrics in a controlled 45° panel test with a 1/4" spacing, consistent differences were observed in heat transferred from cotton, PE/c, and acrylic, with cotton

transferring the maximum amount of heat per gram fabric, and
PE/c fabrics significantly less. These results were shown in
Table 5-4.

Thus, the observations on heat transfer behavior of the
cotton and the PE/c blends include: (1) approximately equal
heat transfers recorded during garment burning tests, and
(2) significantly different heat transfers recorded during
fabric strip burning tests at 45° and horizontal, with spacings
of 1/4 in and 1/2 in, respectively. The maximum differences
are observed in the horizontal test at 1/2 in-spacing, with
cotton fabrics exhibiting greater heat transfer. It would be
desirable to have additional data and, above all, data on the
heat shrinkage characteristics of the fabrics involved, since
significant differences in thermal shrinkage may affect the
fabric-mannequin spacings in the garment burn tests significantly,
and may also influence shrinkage tensions in the framed strip
specimen tests.

5.5.4 Inclined (45°) and Vertical Strips

The M.I.T. group ran numerous burning tests at 45°
orientation, with a 1/2 in spacing. The test environment was
kept at 70°F and 65% RH. Simulant temperatures were
normalized to 37°C, and temperatures measured at 80 μm
(simulated) were used to calculate temperature histories and
modified damage integrals at various skin depths. The simulant
temperatures were measured at a point 16 in from the point of
ignition. Some test results are shown in Table 5-7 for five
cotton fabrics ranging in weight from 1.9 oz/yd^2 to 8.66 oz/yd^2
and the damage reported ranged from 270 μm to 3700 μm. (Recall
that the epidermis thickness may be taken as 80 μm and the
dermis depth runs to 2000 μm.) Thus, it becomes easy to convert
the depth readings of Table 5-7 into characterization of skin
burn damage.

Table 5-7

1/2 INCH SPACING [R25]

Fabric	Weight (oz/yd^2)	Type of Burning	Average Flame Speed (cm/min)	Depth of Damage in Skin (microns)
10 (yellow)	1.90	good	304.8	320
10 (yellow)	1.90	good	304.8	300
10 (yellow)	1.90	good	304.8	270
18	3.69	good	145.0	800
18	3.69	good	139.0	850
18	3.69	good	139.0	750
18	3.69	good	139.0	800
5	3.97	good	113.0	1400
5	3.97	good	113.0	1800
5	3.97	good	114.0	1650
9	7.42	good	67.5	2700
9	7.42	good	65.0	2850
9	7.42	good	66.3	2800
4	8.66	good	67.7	3250
4	8.66	good	69.0	3150
4	8.66	bad	65.0	3700

The effect of fabric weight on burn damage produced
in 45° strip tests at 1/2" spacing for the cotton and PE/C
fabrics is shown in Fig. 5-15. It is seen that the plots
obtained for the two types of fabric are two different
straight lines, with cotton fabric again showing higher
damage as reported for the horizontal burning tests shown
in Fig. 5-14.

The effect of fabric weight in 90° upward burning
tests at 1/2" spacing is shown in Fig. 5-16, where again
results on the two sets of fabric (cotton and PE/C) fall
on separate lines, with cotton producing the higher level
of damage. The blend results show considerable scatter,
but the cotton results fall very close to the linear relationship drawn.

In general, the damage noted in 90° upward burning
is greater than observed in upward burning at 45°, although for the cotton the increase   due to this change
in orientation is small. It appears to be more pronounced
for the blends, particularly in the case of heavier fabrics.

5.5.5 Fabric Combinations

In the case of thermoplastic fabrics, ignition in
combination with a cellulosic was one way of obtaining
more regular burning and reasonably reproducible results.
This was also true of the wool fabric (#20). The cellulosic fabric selected by the M.I.T. group for these strip
tests on fabric combinations was #5, the knit cotton T-
shirt material, 3.97 oz./yd.^{2}. The fabric pairs were pre-
tensioned and fixed in a frame inclined at 45°, and tests
were conducted for both upward and downward burning. The
spacing between fabric and skin-simulant was set at 1/2" for
all "composite" tests. The cotton fabric was placed underneath.

It was found that all the thermoplastic fabrics tested
in this manner melted and shrank ahead of the invisible flame.
Their middle portions sagged and burned less rapidly than

299

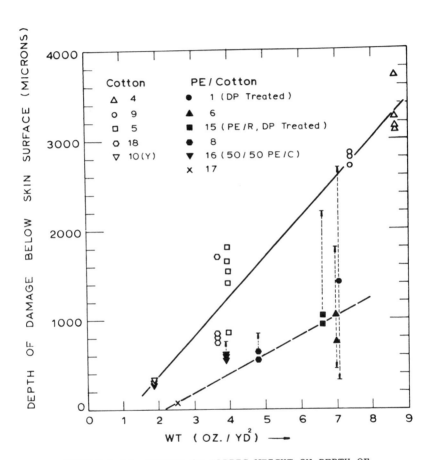

FIGURE 5-15 EFFECT OF FABRIC WEIGHT ON DEPTH OF
DAMAGE (45° UP BURNING, 1/2 INCH SPACING,
SKIN BELOW FABRIC, 65% R.H. AT 70°F) [R25]

300

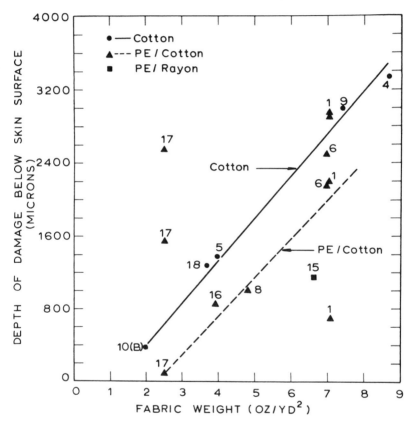

FIGURE 5-16   EFFECT OF FABRIC WEIGHT ON
BURNING INJURY, 90° UP BURNING, 1/2 INCH SPACING   [R25]

the specimen sides. Accordingly, the ignition source in the form of a 1/2" wide strip of computer card was concentrated at the specimen center. In contrast, the wool fabric burned uniformly without sagging, together with the knit cotton backing fabric, although it was observed that the wool appeared to melt slightly ahead of the flame front, leaving a thick char after burning.

The results of the M.I.T. composite burning experiments have been presented in Chapter IV in the discussion of flame propagation velocities. In Table 4-3 information was summarized on fabric composite identification, mode of ignition, type of burning, average flame speed and depth of skin damage with notations on burning behavior of the specimens. The "bad" burns were observed to transfer much less heat than the "good" burns. For the reader's convenience the "good" burn data from Table 4-3 are now repeated in Table 5-8.

It has been noted in Chapter IV that all the composites in this test series burn more slowly than does the cotton knit fabric alone, a result consistent with garment burn data reported in Figure 4-15. It is to be expected that heat transferred from burning of double layers of similar fabrics should be approximately the same as the sum of heats transferred from separate burning of the component layers. This has been shown to hold true for cottons and for PET blends [34]. Chouinard noted that when on the other hand a thermoplastic was burned in combination with a cotton fabric with the thermoplastic underneath, heat from the burning cellulosic caused the thermoplastic to form molten droplets which fell to the sensor surface and extinguished before burning completely. Thus the heat potential of the thermoplastic was not fully realized and the average heat transfer of the combination was only

302

TABLE 5-8

UPWARD 45° BURNING OF WOOL AND THERMOPLASTICS
WITH A COTTON KNIT FABRIC BACKING

(GOOD BURNS ONLY) [R25]

Identification		Weight oz/yd^2	Average Flame Speed cm/min	Permeability (ft/sec)/(lb/ft^2)	Depth of Skin Damage (microns)
Nylon Taffeta #14		1.67	72.6	0.58	1150
Nylon Taffeta #14		1.67	65.0	0.58	1200
Nylon Tricot #12		2.47	58.6	7.4	2500
Nylon Tricot #12		2.47	59.8	7.4	2700
PET Twill #2		2.19	75.0	0.70	900
PET Twill #2		2.19	75.0	0.70	1300
PET D-knit #3		5.83	44.8	5.1	1350
Acrylic J-knit #7		2.64	61.0	2.6	2050
Acrylic J-knit #7		2.64	--	2.6	1850
Acrylic J-knit #7		2.64	--	2.6	1700
Acetate/Nylon	#11	2.67	69.3	0.40	1800
(knit)	#11	2.67	66.3	0.40	1300
Acetate Knit	#13	2.44	78.2	6.8	2050
Acetate Knit	#13	2.44	--		1500
Wool (napped)	#20	6.48	49.2	0.92	3500
	#20	6.48	50.8	0.92	3500
	#20	6.48	--	0.92	3500
Cotton Knit #5 (alone)		3.97	113.0	1.18	1620

slightly greater than that of the cotton alone.

The M.I.T. tests of 45° upward burning of thermoplastics in combination with a cotton underlayer (#5)
show numerous instances (See Table 5-8) where the composite gives off less heat (hence less skin damage) than
does the burn test of the cotton alone. Examples of this
behavior are seen in the Nylon/cotton composite (#14/#5)
PET/cotton composites (#2/#5) & (#3/#5) and, in at least
one test, the acetate/nylon blend/cotton (#11/#5) and
the acetate knit/cotton (#13/#5). On the other hand the
slightly heavier nylon tricot (#12) and the very heavy
wool fabric (#20) transfer significantly more heat than
the cotton alone when burned with the cotton (#5) underlayer.

The M.I.T. group discusses the role of relative
porosity of the two fabrics in the composite suggesting
that melting of the thermoplastic fabric can impregnate
the base (cotton) fabric, "reducing its porosity, resulting in incomplete combustion of the composite."
They reason that thermoplastics (such as #14) with low
porosities tend to seal the underfabric more easily thus
suppressing combustion, while for thermoplastics of
higher porosities (e.g. #12) the change would be smaller,
and more complete combustion would occur with resulting
higher heat transfer.

This reasoning is based on the premise that porosity
is directly related to threads per inch, and implies that
porosity is equivalent to air permeability. We may consider the air permeability of a textile material to be
directly related to geometric porosity. From the values
for permeability of the GIRCFF fabrics measured by the
Factory Mutual Research Group and included in Table 5-8,
it is evident that fabrics #14, #2, and #11 which appear

304

to restrict combustion in combination with the cotton #5, are indeed the least permeable of the group. However, the double knit polyester fabric #3, which is more permeable, also restricts combustion of the composite. The wool fabric #20, which has relatively low air permeability, contributes most to the heat of combustion transferred to the simulant, but in the case of wool, changes in permeability due to melting would clearly be different.

It should be pointed out that fabric air permeability is, in fact, a function of yarn diameters, yarn spacings (or threads per inch), interlace geometry (i.e. knot structure or weave, as well as relative crimp) and mechanical finish (such as needling or fulling). Thus, while the suggestion that reduced permeability of the cotton underlayer is an inter-active variable in the burning of composites may be valid, it might be well to consider also the effect of heat shrinkage in advance of the flame front on the permeability of the thermo-plastic top fabric itself. Finally, it might be well to consider the effect of the melt viscosity of the thermoplastic as it relates to "sealing" of the cotton fabric underneath. Clearly, there is much more to be learned about the combustion of fabric pairs which differ widely in structural and thermophysical properties.

The downward burning at 45° of thermoplastic/cotton composites was examined in an attempt to overcome the stepwise burning which had been observed in the upward burning tests at 45°. The flame propagation speed in downward burn tests was of the order of 10% of that in upward burn tests. These steps are attributable to evasive shrinking of the thermoplastic ahead of the flame to a point of jamming--whereupon the thermoplastic would ignite and

burn out the jammed region. The adjacent virgin fabric
zone would again shrink and thus evade the flame - until
it too could shrink no more due to jamming, - then it
would melt, ignite, and burn. By burning the composites
downward at 45° the M.I.T. group hoped to suppress ex-
cessive fabric shrinkage ahead of the flame, and thus
eliminate this stepwise mode of burning. In fact, downward burning propagated with a regular speed and as was
noted earlier, the thermoplastic melted slightly ahead
of the flame and forming small globules which rested
on the backing material and fed the flame. In this
configuration, the heat transfer and burn damage
resulting from burning of the composites were significantly
higher than for fabric #5 burned downward alone. The
data obtained are shown in Table 5-9. Permeability no
longer seemed to play a role. What was considered important was the observation that the PET fabrics burned
with a regular, luminous and intense flame. The nylons
however melted in front of the flame and tended to flow
towards the specimen center forming a few fuel globules
which caused localized burning, and consequently lower
heat transfer than for the regular, level flame front
in the case of polyester.

5.6 MECHANISM OF HEAT TRANSFER TO SKIN
The M.I.T. study of burn injury resulting from burning
fabrics was extended to cover the various modes of heat
transfer present in the burn tests described earlier in
this Chapter. In general, it was stated that "heat may
be transferred from burning fabric to skin by three
mechanisms: (1) convection of hot gases within the air-
gap towards the colder skin, and conduction through a
thin layer of stagnant gas adjacent to the skin surface;
(2) radiation from the hot gas mass and the hot solids

TABLE 5-9

DOWN BURNING AT 45° OF THERMOPLASTIC

FABRICS OVER A KNIT COTTON FABRIC

Identification	Weight of Top Fabric oz/yd^2	Average Flame Speed cm/min	Permeability (ft/sec)/(lb/ft^2)	Depth of Damage Microns
Cotton Knit #5 (alone)	3.97	10.9	1.18	1850
Nylon Taffeta #14	1.67	7.8	0.58	2700
Nylon Tricot #12	2.47	6.7	7.4	2100
PET Twill #2	2.19	8.6	0.70	3300
PET D-knit #3	5.83	5.2	5.1	3500

(fabric and glowing char), and (3) condensation of steam
and pyrolysis products at sufficiently small spacing to
cause flame quenching. For pure thermoplastic fabrics, and
their blends, an additional mechanism, melt-dripping becomes
important."

The M.I.T. study of burn injury was limited to 100%
cotton fabrics and therefore the role of melt-dripping was
not studied. (This was, however, an important part of the
Chouinard study which was discussed earlier in this chapter.)
The M.I.T. experiments on the cotton fabrics required direct
measurement of total heat flux "q_t" on a skin simulant surface,
as well as measurement of the temperature distribution within
the skin-fabric air gap, as a function of time. The flux was
measured with a circular foil radiometer and temperature (T)
measurements were made by means of fine thermocouples. In
addition to the total flux, the conductive heat flux q_c was
calculated on the basis of the relationship

$$q_c = k_g \left(\frac{dT}{dx}\right) \text{ at the surface} \tag{5.4}$$

where k_g is the thermal conductivity of stagnant air adjacent
to the skin.

The heat flux due to radiation from the hot fabric and
char, q_r, is determined from the relationship

$$q_r = 1/2 \int_{-1}^{1} \varepsilon_f \sigma T_f^4 \ d(\sin\psi) \tag{5.5}$$

where ε_f is the emissivity of the hot fabric, σ is the
Stefan-Boltzmann Constant, T_f is the absolute temperature
of the fabric, and ψ is the angle between the line joining
a point on the fabric and the radiometer located in the
skin surface, and the normal from the simulant surface at

the radiometer location.

The heat flux due to condensation of pyrolysis products, q_p, was not measured directly, but whenever condensation was observed visually, its contribution to the total flux was estimated by differences.

Two cotton fabrics were studied in this phase, #10 (blue), and #18. Burn tests were conducted in the horizontal and in the 90° configurations, with 1/2 inch spacing. For 90° orientation, a series of spacings (1/2", 3/8", 1/4" and 1/8")were studied.

5.6.1 Horizontal Samples

Figure 5-17 shows the total surface heat flux and the temperature distribution in the fabric-skin gap for horizontal burning of fabric #10 at 1/2 inch spacing. The temperature readings at the skin surface, and at a point 0.06 cm above it were used to calculate the heat flux q_c due to conduction. This temperature peaks 39 seconds after ignition, as does the total heat flux q_t measured by means of the foil radiometer. The maximum of the total flux, reached just before the fabric has burned out is 0.26 cal/cm^2-sec as indicated by the drop in the fabric temperature at 1.27 cm above the skin ($T_{1.27}$).

The data of Fig. 5-17 is replotted in Fig. 5-18 so as to emphasize the distribution of temperatures over the fabric skin gap at various times. It is seen that maximum gas temperature occurs 0.9 cm from the skin or 0.4 cm below the fabric surface and heat is accordingly conducted both to the fabric and to the skin from this region of combustion. It is noted that the maximum temperature reaches 910°C at 39 sec.

Because of difficulty in measuring fabric temperature at the 1.27 cm location of the thermocouple, values of q_r, calculated by means of equation (5.5) were found

309

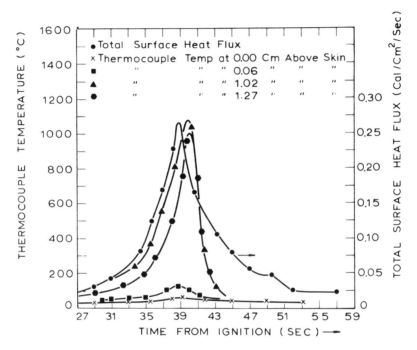

FIGURE 5-17   TOTAL SURFACE HEAT FLUX AND
TEMPERATURE DISTRIBUTION IN THE AIR GAP,
FABRIC 10(B), HORIZONTAL, 1/2 INCH SPACING   [R25]

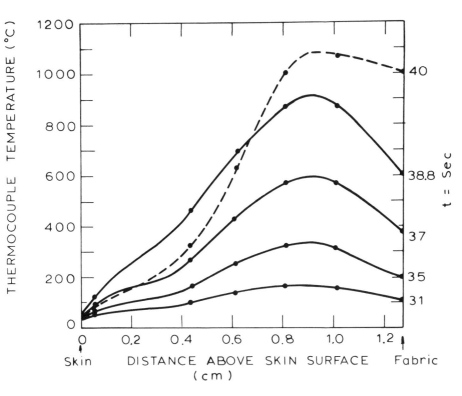

FIGURE 5-18 TEMPERATURE DISTRIBUTION IN FABRIC-SKIN
AIR GAP AT VARIOUS TIMES DURING FLAME SPREAD, FABRIC
#10(b) HORIZONTAL, 1/2 INCH SPACING [R25]

to exceed measured values of total heat flux, q_t. This led to revision of the technique for calculation of the radiant flux q_r. The comparative results for q_t, q_r (corrected) and q_c at three different times after ignition are shown in Table 5-10. Radiation from the hot fabric is seen to go as high as 65% of the total flux. In comparison of columns 4 and 5 one notes good agreement in the values of q_r calculated directly and by differences except at 37 seconds.

Similar test data has been reported for the heavier all cotton fabric #18 (3.69 oz/yd^2), and plots similar to Figures 5-17 and 5-18 have been reported. They showed the maximum temperature between skin and burning fabric #18 to rise to a maximum of 940°C at 0.9 cm above the skin (compared to 910°C at 0.9 cm above the skin for lighter fabric #10). This higher temperature gradient near the skin was thought to provide a higher conductive contribution to the total heat flux for the heavier #18.

In the absence of condensation heat transfer, the radiant flux q_r was assumed equal to the difference between total flux and conductive flux, and based on the q_r thus calculated, the inference was drawn that the radiative component of the total flux was less for the heavier fabric #18, i.e. about 50% as compared to 65% for fabric #10. The total heat flux from fabric #18 reached a level twice that observed for fabric #10. However, in repeat tests this relatively high total flux difference between #18 and #10 was considerably reduced, an occurrence which was attributed to char formation and to the effect of sagging or falling of char (during horizontal burning) on air recirculation and hence on temperature distribution in the skin-fabric air gap.

Finally, the total thermal dosage to the skin during the horizontal burn tests was obtained by graphical integration of the area of the q_t vs. time curve such as the

Table 5-10

TOTAL, CONDUCTIVE AND RADIATIVE HEAT FLUXES TO SKIN DURING
HORIZONTAL BURNING OF FABRIC #10 AT 1/2" SPACING (1.9 OZ/YD^2)

t sec	q_t cal/cm^2 sec	q_c cal/cm^2 sec	q_r cal/cm^2 sec	$(q_t - q_c)$ cal/cm^2 sec	$(q_t - q_c)/q_t$ %
37.0	0.168	0.060	0.135	0.108	64
38.8	0.265	0.092	0.174	0.173	65
40.0	0.200	0.066	0.140	0.134	67

one shown in Fig. 5-17. Similar data were obtained and
integrated for repeat tests for both fabrics. The results
reported in Table 5-11 indicate that variations in burning
conditions such as char formation and fabric sagging (which
were observed in the repeat tests) can significantly affect
maximum heat flux readings and can similarly affect the time
integrated value of thermal dosage. We also note that the
thermal dosage averages about 3 cal/cm^2 for the cotton batiste
#10 (2 oz) and 7.5 cal/cm^2 for the cotton flannel #18
(3.7 oz).

Table 5-11

THERMAL DOSAGE TO SKIN IN HORIZONTAL BURNING
AT 1/2" (REPEAT TESTS)

Fabric	Weight (oz/yd^2)	Maximum Flux cal/cm^2 sec	Thermal Dosage cal/cm^2
#10	1.98	0.33	3.67
	1.98	0.26	2.55
#18	3.69	0.53	8.46
	3.69	0.35	6.58

5.6.2 Vertical Samples

The upward burning (90°) of cotton fabrics #10 and #18
was also studied by the M.I.T. group and results for total heat
flux and temperature distribution as a function of time are
shown in Fig. 5-19. As expected, the velocity of flame
propagation was much higher for the vertical burn. The maximum flux was 0.77 cal/cm^2 sec for the vertical burning of the

314

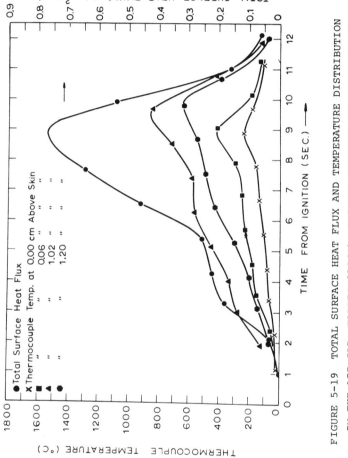

FIGURE 5-19 TOTAL SURFACE HEAT FLUX AND TEMPERATURE DISTRIBUTION
IN THE AIR GAP, FABRIC #10(B), 90° UP, 1/2 INCH SPACING [R25]

315

cotton batiste #10, more than twice that shown in Table 5-11
for the horizontal test, and the total thermal dosage suffered
by the skin was also considerably higher for the vertical burn.
The flame between the fabric and skin was 5 to 6 cm long, and
it appeared that the fabric burn out approached the radiometer
location about one second after the maximum heat flux was
observed at that location. The pyrolysis region of the fabric
(as indicated by blackening) appeared to extend two to three
inches above the burned out region. Finally, the maximum
temperature of 1200° was observed to occur 0.7 cm from the skin
(as compared to 910°C at 0.9 cm for the horizontal burn).

These characteristics of the vertical burn should contribute to higher conductive fluxes (via gas temperature
gradients) and higher radiative fluxes (via greater areas of
hot fabric ahead of the flame, and hotter gases). The total
flux readings shown in Fig. 5-19 and the values for flux
components shown in Table 5-12 confirm that this is, in fact,
the case.

Table 5-12

COMPARISON OF TOTAL CONDUCTIVE AND RADIATIVE

HEAT FLUXES TO SKIN

(FABRIC #10 VERTICAL UPWARD BURN AT 1/2")

Time	q_t Cal/cm² sec	q_c Cal/cm² sec	$q_r = q_t - q_c$ Cal/cm² sec	q_r/q_t %
5	0.239	0.149	0.09	38
6	0.345	0.183	0.162	47
7	0.535	0.171	0.364	68
8.9*	0.766	0.341	0.425	55
9.9	0.537	0.109	0.428	80
10.5	0.340	0.040	0.300	88

*time of maximum flux

316

Table 5-12 shows that the percentage contribution of radiation to total flus reaches about 55% at the time when maximum total flux occurs. One second thereafter, when the burned out region of the fabric moves over the radiometer, it increases to 80%. Still one second later as the total flux continues to drop, the radiative contribution rises to 88%. Beyond this time, heat transfer is essentially radiative.

Table 5-12 has heuristic value beyond the specifics of fabric #10, for it demonstrates the history of a burn transverse over the skin for a typical lightweight, flammable fabric. It shows the development of total heat flux as it may affect skin damage and it characterizes the shift from conductive to radiative heat transfer to the skin during the course of the burn.

The MIT group conducted similar vertical burns with the heavier cotton flannel #18 and reported data comparable to those shown in Fig. 5-19 and in Table 5-12. The location of maximum gas temperatures in the skin fabric gap was the same and the maximum gas temperature itself was similar (at 1160°C) for the heavier fabric #18 and for the lighter #10. The radiation equation was used to calculate q_r (assuming cloth emissivity of 0.8) and the values obtained were compared with q_r values obtained as the difference between total flux q_t and its conductive component, q_c. The q_r values determined by these two methods agreed at a time a few seconds after the occurrence of maximum total heat flux. At times prior to this, when the flame region straddled the radiometer position in the skin simulant, the radiative contribution of the hot gases would be expected to provide a significant portion of the overall radiative heat flux. The data shown in Table 5-13 indicates that this in fact does occur. On the other hand, a few seconds after maximum total flux is reached, the hot fabric component of the radiant flux is

317

dominant, and the radiative flux far exceeds the conductive flux.

Table 5-13

COMPARISON OF HEAT FLUXES TO SKIN IN BURN TEST OF

FABRIC #18   90° UPWARD AT 1/2"

Time (sec)	Total Flux q_t (cal/cm^2 sec)	Conductive q_c (cal/cm^2 sec)	Radiative Total $q_r = q_t - q_c$ (cal/cm^2 sec)	Radiative Hot Fabric (cal/cm^2 sec)	q_r/q %
4	0.200	0.123	0.077	------	39
10	0.355	0.136	0.219	0.120	62
14	0.580	0.156	0.424	------	73
18*	0.805	0.377	0.428	0.350	54
20	0.625	0.178	0.447	0.470	72
21	0.480	0.102	0.378	0.320	79

*Time of maximum heat flux

These observations are made on the basis of comparison of the last three columns in Table 5-13. It is also noted in the second column that the total heat flux during the vertical burn test of fabric #18 rises well above 0.2 cal/cm^2 sec for over 17 secs. Reference to Stoll's data given in Fig. 5-6 indicates that this dosage far exceeds that needed to inflict second degree burns. Even for the much lighter fabric #10 the heat flux data of Table 5-12 show a total flux above 0.25 cal/cm^2 sec for more than five seconds, namely a situation which just about corresponds to a second degree burn condition.

In conclusion, one notes that in vertical upward burning of these two cotton fabrics (#10 and #18), nearly same values of total, conductive and radiative heat fluxes are attained, while the velocity of flame propagation for the lighter fabric is about twice that of the heavier fabric. The total thermal dosage inflicted on the skin by the slower burning fabric is, therefore, about twice that imposed by the faster burning fabric.

5.6.3 Effects of Fabric-Skin Spacing

The M.I.T. group went on to study the effect of fabric skin gap on the heat flux components during burn tests of cotton fabrics, at 1/2, 3/8, 1/4 and 1/8" spacings. The results of the upward vertical burn tests at 1/2" spacing have been discussed above. Results for the other spacings are summarized below.

As for the 1/2" spacing tests, the data were reported in three forms: (1) in graphs of total heat flux and temperature history in the air gap as in Figs. 5-17 and 5-19, (2) plots of temperature distribution as a function of distance from the skin as in Fig. 5-18, and (3) in tables comparing heat flux components at various times after ignition, as in Tables 5-11 and 5-12. Rather than reproducing here all the graphs and tables, the findings will be discussed in a general way.

The summary of test data given in Table 5-14 includes results for various spacings in vertical upward burning of fabrics #10 and #18: the maximum total heat flux, the maximum skin (simulant) temperatures achieved, the corresponding percent radiative (q_r) and condensing (q_p) component of that maximum flux, the maximum flame temperature in the gap, the flame velocity, the thermal dosage, and the calculated

319

Table 5-14

COMPARISON BETWEEN HEAT TRANSFER AND THERMAL INJURY

Fabric #	Orientation	Spacing (inch)	$q_{t,max}$ (cal/cm^2 sec)	$T_{s,max}^{f}$ (°C)	$(q_p+q_r)/q_t$ (%)	$T_{x,max}^{ff}$ (°C)	V av. (cm/min)	Q_t (cal/cm^2)	Skin Damage (microns)
10	Horizontal	1/2	0.30	100	65	910°	20.0	3.7	80
18	Horizontal	1/2	0.44	100	54	940°	13.5	6.6	500
10	90° Up	1/2	0.766	240	55	1200°	254.0	4.6	390
18	90° Up	1/2	0.805	240	54	1160°	150.0	10.4	1200
18	90° Up	3/8	0.780	240	30	1150°	105.0	8.4	1150
13	90° Up	1/4	0.680	270	23	660°	70.0	9.8	900
18	90° Up	1/8	0.460	130	63	500°	25.0	14.6	1850

$^{f}T_s$ = Skin simulant temperature

$^{ff}T_x$ = Flame temperature in air gap

depth of skin damage. Two plots of total flux vs. time after
ignition and temperature distribution in the air gap vs. time
are shown in Figs. 5-20 and 5-21 for fabric #18 burned vertically with 3/8" gap and 1/8" gap respectively. A comparison
of Figs. 5-17 for horizontal burning and 5-19 for vertical
burning of fabric #10 (both tests at 1/2" spacing) shows the
skewing tendency of the vertical burn data. In other words,
horizontal burning gives a more or less symmetric heat history,
whereas in vertical burning the major part of the thermal
dosage is imposed before the time of maximum flux. Figure 5-20
for the vertical burning of fabric #18 with a 3/8" spacing,
shows an even greater skewing of the temperature history viewed
by the skin. In the case of the 1/8" spacing for vertical
burning of fabric #18, which is shown in Fig. 5-21, a skewed
temperature time history is also noted but, in addition,
secondary peaks of fabric temperature and heat flux, corresponding to the afterglow of char become evident in this case.

The total thermal dosage for the various orientations
and spacings of fabric #10 and fabric #18 are shown in
Table 5-14. It is seen that the maximum dosage is suffered
by the skin in the 1/8" spacing vertical burn test of #18.
The values given in Table 5-14 were obtained by means of
graphical integration of Figs. 5-17, 5-19, 5-20, and 5-21,
and corresponding graphs not reported here.

More should be said about the value of condensation
products heat flux listed in Table 5-14 as q_p. In burn tests
of fabric #18 at 1/2" and 3/8" spacings, no condensation
products were observed. However, in tests with air gaps of
1/4" and 1/8", no flame was observed between fabric and skin
and the simulant surface was found to be coated with a tarry
material. The data of heat flux components for the 1/2" and

321

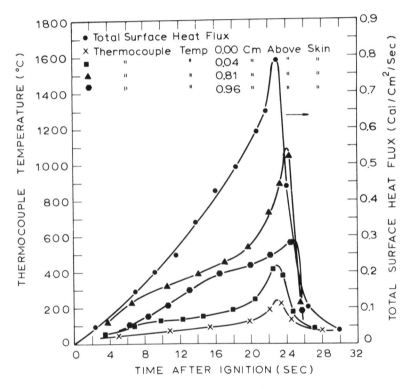

FIGURE 5-20   TOTAL SURFACE HEAT FLUX AND
TEMPERATURE DISTRIBUTION IN THE AIR GAP, FABRIC #18,
90° UP, 3/8 INCH SPACING   [25]

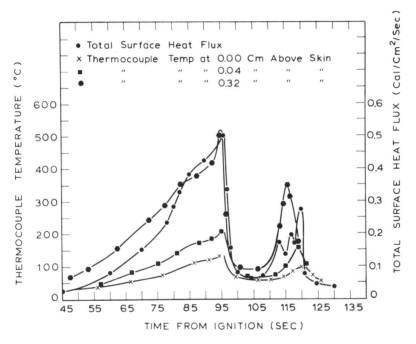

FIGURE 5-21 TOTAL SURFACE HEAT FLUX AND TEMPERATURE
DISTRIBUTION IN THE AIR GAP FABRIC #18, 90° UP,
1/8 INCH SPACING [R25]

3/8" spacing tests were then compared with those reported for
the 1/4" and 1/8" tests. It was noted that the difference
between the total heat flux q_t and the conductive flux q_c
could not be taken as an estimate of the radiative flux q_r
for the 1/4" - 1/8" tests, for it was clear that a condensative flux was also present. Thus, $(q_t - q_c)$ was taken as a
determination of the sum $(q_r + q_p)$. It was then assumed that
at the time of maximum total flux, $q_{t,max}$, the contribution of
the condensation flux q_p was negligible and thus at this
moment, that $(q_t - q_c)$ was equal to q_r. Further, it was
assumed that the radiative flux q_r behaved in the 1/4" and
1/8" tests as it had in the 1/2" and 3/8" tests, reaching a
maximum near the time of maximum total flux, and diminishing
linearly in the time period before that. This assumption
permitted the determination of the q components in the earlier
stages of 1/4" burn tests, giving an average value of
0.15 cal/cm² sec, or <u>30 to 40%</u> of the total heat transfer in
the time period from 34 to 43 secs after ignition. In similar
calculations from data obtained during burnings of fabric #18
with a 1/8" spacer, it was shown that the heat flux due to
condensation of pyrolysis products accounted for 30-35% of
the total heat transfer during the critical time period.

5.6.4 Summary

The M.I.T. report has provided in Table 5-14 a summary of
results for heat transfer measurements between burning cotton
fabrics #10 and #18 and an adjacent skin simulant. Their
comment follows:

"Thermal injury, being a rate process and thus a function
of temperature history of the tissue, and the time for which
these temperatures are above a minimum injurious level, a
qualitative measure for fabric burn tests of these two

quantities can be obtained from values of $T_{s,max}$* and V_{av}**
the inverse of the latter quantity being a measure of the
exposure time.

For horizontal burning, $T_{s,max}$ are equal for both
fabrics, but the values of exposure time (inverse of V_{av}),
Q_t, and depth of damage for the heavier fabric #18 are higher
than those for fabric #10. Similar behavior is found for 90°
up burning at $\delta = 1/2$". A comparison of the horizontal and
90° up orientation at $\delta = 1/2$" for, say, fabric #18, shows
that even though the exposure times are an order of magnitude
smaller for the latter mode, $T_{s,max}$, Q_t and depth of damage
are much higher.

The effect of spacing in 90° burning of fabric #18
can be seen by the results in the last four rows of Table 5-14.
The flame speeds decrease with decrease in δ; both Q_t and
depth of damage, however, go through a minimum. The increase
in Q_t and depth of damage below a critical spacing is due to
the effect of condensation of pyrolysis products. $T_{s,max}$ for
$\delta = 1/4$" appears to be the highest.

Finally, a comparison between the total thermal dose,
Q_t, and depth of damage in the last two columns shows a
striking similarity in behavior. Depth of skin damage appears
to be a direct function of the total heat delivered to skin.
Figure 5-9b is a plot of depth of skin damage vs. total
thermal dose, Q_t, to skin. The data appear to follow
approximately a straight line correlation (as shown by M.I.T.
and GRI results). Below a Q_t of about 4 cal/cm^2 the damage
depth is seen to attain a nearly constant value of 80 microns
down to a Q_t of = 2.5 cal/cm^2. The constancy of damage at 80
microns from Q_t = 2.5 to 4 is due to the energy barrier for
damage at the dermal-epidermal junction. Dermis being more
sensitive to damage than the epidermis, a certain excess
energy is required for the damage to propagate into dermis.

---

* skin temperature

** average flame velocity

5.7 OVERVIEW OF STUDIES OF BURN INJURIES

The local injury potential of burning fabrics is determined
by the heat flux and by the total thermal dosage absorbed by
the skin. Heat flux from a burning fabric is not a material
property; it is a rate of heat transfer which depends on the
conditions under which burning occurs. Factors of importance
are the availability of oxygen, the orientation of the fabric
and skin, the direction of burning for a given fabric-skin
orientation, the spacing between fabric and skin, the nature of
the fiber and its thermal behavior (i.e. shrinkage and melting),
the juxtaposition of different fabrics in a composite system
undergoing burning.

In a study related to the GIRCFF program, Yeh[18] determined the heat of combustion in oxygen for a variety of
commercial fabrics of varying fiber content and found that a
gram of fabric burned (in O_2) gave off from 3700 to 7000
calories, with cotton at the low end and nylon and acrylic
at the high end of this range. When the same fabrics were
burned in air (with 21% O_2), the total heat given off ranged
from 2200 to 4100 calories per gram, with acrylic still highest,
but with polyester now lowest and cotton about midway. The
fiber content and, in a few cases, the structure determined
the efficiency of combustion. Tables 5-15a, b and 5-16 give
the heat of combustion in oxygen and the heat of burning in air
for GIRCFF and non-GIRCFF fabrics as determined by Yeh[18] and
also the heat transferred to skin simulants under various testing
conditions, as reported by others. These tables will be
discussed further below.

The fact that Yeh's data on heat of burning are reported
in terms of cal/g suggests that if one burns 10 grams of fiber
in one experiment as against 5 grams in another, the total heat
output will be in a ratio of 2:1. Or, if one burns comparable
areas of a 10 oz/yd^2 fabric in comparison with a 5 oz/yd^2

326

Table 5-15a

HEAT OF COMBUSTION, HEAT OF BURNING AND HEAT TRANSFERRED TO SKIN UNDER VARIOUS CONDITIONS

1. B.B.P. = Base Burn Potential defined by Chouinard (34) as Heat Transferred/Fabric Weight
2. V = Vertical Strip
3. 45° = Strip placed in a 45° plane
4. G = Garment burned on a stationary mannequin

Fabric Material	Weight oz/yd²	Heat Combustion in O_2, H_c (18) cal/g	Heat of Burning in Air H_b (18) cal/g	cal/cm²	Orient.	Spacing in	Heat Transfer. $H_{b,t}$ cal/cm²	cal/g	Rel. Heat Transfer. $H_{b,t}/H_b$ %	B.B.P.[1] cal/cm²	oz/yd²	Reference
Wool	-	4950	-	-	V up[2]	0.5	-	1200	-	-	-	Webster (50)
Cotton	-	4180	-	-	V up	0.5	-	1200	-	-	-	Webster (50)
Cotton	-	-	-	-	V up	1.5 max	-	-	-	-	-	Alvares (51)
--	-	-	-	-	V up	0.5 max	-	-	-	-	-	Brown (52)
Cotton	-	3688	3315	-	45° up[3]	0.25	-	500	15		1.7	Chouinard (34)
Nylon	-	6926	3715	-	45° up	0.25	-	450	12		1.5	"
Wool	-	-	-	-	45° up	0.25	-	-	-		1.3	"
PE/C 65/35	-	4907	2776	-	45° up	0.25	-	390	14		1.3	"
Acrylic	-	7020	4118	-	45° up	0.25	-	350	9		1.2	"
Polyester	-	5255	2188	-	45° up	0.25	-	240	11		0.8	"
Cotton	3.2	-	3315	36	G[4]	-	8.0	740	22		2.5	"
Cotton	7.0	-	3315	79	G	-	14.6	620	19		2.1	"
PE/C 65/35	2.8	-	2776	26	G	-	7.3	770	28		2.6	"
PE/C	5.3	-	2776	50	G	-	11.3	630	23		2.1	"
Acrylic	3.3	-	4118	46	G	-	7.1	640	15		2.2	"
Acrylic	6.6	-	4118	92	G	-	12.3	550	13		1.9	"
Wool	3.5	-	-	-	G	-	5.9	500	-		1.7	"
Wool	7.5	-	-	-	G	-	7.9	310	-		1.1	"

Table 5-15b

HEAT OF COMBUSTION, HEAT OF BURNING AND HEAT TRANSFERRED TO SKIN UNDER VARIOUS CONDITIONS

Fabric #	Type	Weight oz/yd²	Heat of Burning in Air H_{ba} (18)		Orient.	Spacing inches	Heat Transferred H_{bt}		Base Burn Potential (BBP) cal/cm²//oz/yd²	Rel.Heat Trans. H_{bt}/H_{ba}	Ref.
			cal/g	cal/cm²			cal/cm²	cal/g			
Cotton-10	Batiste	2.0	3120	21	G*	--	10	1500	5.0	48	(R28)
Cotton-18	Flannel	3.7	3270	41	G	--	17	1290	4.6	41	"
Cotton- 4	Denim	8.7	3200	96	G	--	23	780	2.6	24	"
PEC/C- 17	Batiste	2.5	2970	27	G	--	12	1320	4.8	44	"
PE/C- 16	Shirting	3.9	2780	36	G	--	13	990	3.3	36	"
PE/C - 8	Jersey	4.8	3160	51	G	--	16	990	3.3	31	"
PE/C - 6	Twill	7.0	2790	67	G	--	21	870	3.0	31	"
PE/C - 1	DP Twill	7.1	2660	63	G	--	20	840	2.8	32	"
Cotton-10	Batiste	2.0	3120	20	H**	1/2	3.7	560	1.9	19	(R25)
Cotton-18	Flannel	3.7	3270	41	H	1/2	6.6	500	1.5	16	"
Cotton-10	Batiste	2.0	3120	20	V(up)***	1/2	4.6	690	2.3	23	"
Cotton-18	Flannel	3.7	3270	41	V(up)	1/2	10.4	800	2.8	25	"
Cotton-18	Flannel	3.7	3270	41	V(up)	3/8	8.4	650	2.3	20	"
Cotton-18	Flannel	3.7	3270	41	V(up)	1/4	9.8	750	2.7	24	"
Cotton-18	Flannel	3.7	3270	41	V(up)	1/8	14.6	1100	4.0	36	"

*   G=Garment
** H=Horizontal Strip
*** V=Vertical Strip

Table 5-16

ESTIMATE OF HEAT TRANSFERRED TO SKIN FROM MIT'S DEPTH OF BURN MEASUREMENTS
AND GRI'S PATHOLOGICAL STUDIES IN RATS (FIG. 5-9a)*

Fabric #	Type	Weight oz/yd^2	Heat of Burning H$_b$ [18] cal/g	cal/cm^2	Spacing Inches	Depth of Burn at Different Orientations H μm up	45°up μm	Vup μm	Range of Heat Transfer for 3 Orient. H$_{b,t}$ cal/cm^2	Reference
Cotton-10	Batiste	2.0	3120	21	1/2	80	300	390	5- 6	[R-25]
Cotton-18	Flannel	3.7	3270	41	1/2	500	800	1280	6-10	"
Cotton- 5	Jersey	4.0	3160	44	1/2	650	1600	1380	7-10	"
Cotton- 9	Terry	7.4	--	--	1/2	1650	2800	3000	13->14	"
Cotton- 4	Denim	8.7	3200	96	1/2	2100	3200	3350	13->14	"
PE/C -17	Batiste	2.5	2970	25	1/2	80	80	80	5- 5	"
PE/C -16	Shirtin	3.9	2780	36	1/2	80	500	850	5- 8	"
PE/C - 8	Jersey	4.8	3160	51	1/2	550	680	1000	7- 9	"
PE/C -15	DP Plain	6.6	--	54	1/2	1900	700	1150	12-10	"
PE/C - 6	Twill	7.0	2790	67	1/2	600	900	2150	7-13	"
PE/C - 1	DP Twill	7.1	2660	63	1/2	2800	1700	2200	>14	"

*Rough estimates from Figure 5-9a which gives heat input vs. burn depth for white rat skin and fabrics burning vertically upwards at 1.5" from skin.

material of the same fiber composition, the thermal dosage
received by a nearby sensor will be in the ratio of 2:1. A
linear relationship between thermal dosage (transferred to a
skin simulant from a burning fabric) and weight per unit area
of that fabric has been observed for practically all the cotton
fabrics evaluated in the GIRCFF program--whether in laboratory specimen form, or in single-garment mannequin tests.
This relationship holds also for all the polyester/cotton
blend fabrics tested. Some of Chouinard's [34] specimen burn
data show that it holds for acrylics and for wool as well.
(See Fig. 5-10.)

Chouinard measured the heat transferred from 45° upward-
burning fabrics spaced 0.25" from a sensor board, and reported
that the efficiency of heat transfer to skin for selected
fabrics ranged from 9% to 15% of the heat of burning, with
cotton and PE/c blend exhibiting maximum efficiency, and the
acrylics and polyester the lowest (see Table 5-15a). An
important parameter listed by Chouinard was the mixed dimension
quantity of heat transferred per unit weight expressed in
$cal/cm^2/oz/yd^2$ which is termed a "Base Burn Potential". This
base potential ranged from a low of about 0.8 for polyester
to a high of 1.7 for cotton. One need only multiply the base
burn potential by the fabric weight (in oz/yd^2, the technological dimensions used in commerce) to obtain the likely heat
dosage received by the skin (or a simulant material) under the
given condition of burning. Chouinard's data are summarized
in Table 5-15a for fabric specimen burn tests over a sensor
board, and also for garment burn tests on an instrumented
mannequin.

The base burn potential reported by Chouinard for various
materials burned in garment tests is considerably higher than
observed for the same materials in the 45° upward specimen
burn test. Garments made from the cottons and PE/c ranged from
2.1 to 2.6, the acrylics from 1.9 to 2.2, and the wools from

1.1 to 1.7. The thermoplastic 100% nylons and 100% polyesters showed no Base Burn Potential in these garment tests for the simple reason that they did not propagate flame over the mannequin sensors and thus did not develop any significant heat flux. However, when the thermoplastic garments were worn in conjunction with cotton garments the assembly did propagate flame and the Base Burn Potential was 1.2 for a polyester dress worn over a cotton slip and 1.5 for nylon worn over cotton. A reversal of fabrics in these combinations, (i.e., putting the thermoplastic inside instead of outside) brings the Base Burn Potential down to about 1.0 and 1.3 for the polyester/cotton combination and the nylon/cotton combination, respectively.

What does this mean in terms of burn injury? If on the basis of Chouinard's experiments we take 2.3 as the Base Burn Potential for cottons and PE/C, and note that in the GIRCFF series fabric weights range from 2.0 to 8.7 oz/yd^2, it follows that the resulting thermal dosages in garment mannequin tests would run from about 4 to over 20 cal/cm^2. It has been pointed out previously that 2 cal/cm^2 is enough for a 2nd degree burn and that above 4 cal/cm^2 one encounters 3rd degree burns and worse. Thus severe burn injury can be expected from sustained burning of all the cellulosic and cellulosic blend fabrics tested.

The picture becomes darker when we examine the results of garment mannequin tests reported by GRI [R28]. In Table 5-15b, the Base Burn Potentials of a number of GIRCFF fabrics have been calculated from the GRI data, and a similarity of results to those Chouinard has been noted for the heaviest cotton and PE/C blends (7 to 8.7 oz/yd^2). In the case of the lighter weight GIRCFF fabrics tested in garment form by GRI Base Burn Potentials were significantly higher than the values reported by Chouinard for cotton and PE/C fabrics. This higher level of heat transfer is also seen in Table 5-15b in the column of heat dosage relative to heat of burning where percentages range from 24% to 48% (vs. 19% to 28% reported by Chouinard). This higher percentage represents, in effect, a higher efficiency of combustion and/or of heat transfer in the specific configurations of burning.

Finally in burning specimen-sensor tests on the GIRCFF fabrics, the M.I.T. Fuels Research Laboratory has reported thermal dosages from $\underline{3.7}$ to $\underline{14.6}$ cal/cm^2, depending on fabric orientation and spacing. These data reflect high efficiencies of burning and heat transfer, namely $\underline{16}$ to $\underline{36\%}$ for two cotton fabrics tested under varying conditions of burning, i.e., spacing and orientation.

Thermal dosage for other fabrics under M.I.T.'s testing conditions can be estimated as shown in Table 5-16. Using the pathological correlation provided in Fig. 5-9, the M.I.T. burn depth data are transferred into rough estimates of thermal dosage at the skin simulant. (It should be noted that the conditions of the MIT testing and those of Fig. 5-9a are different.) The range of these estimates is 5 to more than 14 cal/cm^2, i.e., high enough to cause 3rd degree burns. Finally it will be noted that the rough estimates of heat dosage shown in Table 5-16 compare reasonably well with the MIT direct reportings of heat dosage as shown in Table 5-15.b for fabrics #10 and #18.

In summary, the results of specimen burns over sensor boards and of mannequin garment burn tests reveal that from $\underline{10}$ to $\underline{50\%}$ of a fabrics's heat of burning is transferred to the adjacent skin (simulant). With fabric heats of burning ranging from 2200 to 4100 cal/g or from 20 to 90 cal/cm^2, there is little likelihood that the wearer of a garment made from fabrics such as those tested in the GIRCFF program and shown to propagate flame, could escape extensive burn injury if the garment were accidentally ignited. The thermoplastic fabrics do not propagate flame when tested in single garments, but when tested in combination with a cellulosic garment, they do burn and do transfer heat in a sufficient quantity to cause significant human injury.

The rest is a matter of differences in the extent or depth of injury which the human skin will suffer when exposed to varying conditions of fabric burning. Still, if the mechanism of heat transfer could be sufficiently understood, and perhaps sufficiently controlled, it might be possible to identify conditions which would discourage, and even suppress, the propagation of flames. Further, it can be pointed out that in a textile fabric, the phenomenon of flame propagation is likewise dependent on heat transfer--in this case, transfer from the combustion of pyrolysis products to unburnt portions of the fabric.

The M.I.T. studies on heat transfer from burning fabrics have dealt with various conditions which affect flame propagation as well as heat transfer to an adjacent skin simulant. Their data are comparable to results reported by Chouinard and by Arnold, as indicated in the Summary presented in Table 5-17. As pointed out earlier, Chouinard's 45° upward specimen-burn tests transferred the least heat (9% to 15% of the heat of burning in air) to the parallel sensing board spaced at 0.25". In garment burn tests, Chouinard reported increased percentages, ranging from 13% to 28%. GRI garment tests on the GIRCFF fabrics using essentially the same instrumentation as Chouinard, record heat transfer efficiencies ranging from 24% to 48% for all cotton and for PE/c fabrics.

The M.I.T. tests did not include garment burns, but concentrated on local effects during fabric specimen burns in proximity to a sensor board (serving as skin simulant).

Table 5-17

COMPARISON OF BURNING AND HEAT TRANSFER EFFICIENCIES*

Investigator	Range of Efficiences	Type of Test	Fabrics Tested
Chouinard [34]	9-15%	45° spacing @ 0.25"	Cotton,PE/C, thermoplastic
Chouinard [34]	13-28%	Garment- mannequin	Cotton, PE/C thermoplastic
GRI Arnold [R 28]	24-48%	Garment- mannequin	Cotton, PE/C
MIT Williams [R 25]	16-19%	Horiz spacing @ 0.5"	Cotton
MIT Williams [R 25]	20-36%	Vertical spacing @ 0.12 - 0.5"	Cotton

*Efficiency ≡ Heat Transferred to skin/ Heat of burning in air

Horizontal burnings with fabric 0.5" above the skin simulant evidenced heat transfer efficiencies of 16% to 19% for two GIRCFF cotton fabrics. When the fabric was burned in a vertical position upward, the transfer efficiencies jumped to 20% to 25% for spacings of 0.25" to 0.5", and to 36% for the 0.125" spacing.

The factors influencing total thermal dosage and injury to the skin (simulant) were specified by the M.I.T. investigators as: (a) radiation from hot gases and fabric, (b) convection and conduction through the gas phase, (c) condensation of steam and pyrolysis products, and (d) exposure time, which is inversely proportional to flame speed. They showed that in vertical burnings, average flame speeds increased linearly as the fabric-simulant spacing was increased from 0.125" to 0.50". Thermal dosage, on the other hand, was maximum at the 0.125" spacing, fell to a minimum at 0.25", then rose to another peak just below 0.50" spacing. It was concluded that for spacings below 9/32" there is not enough air for complete combustion and some flame quenching occurs; consequently, the temperature of the hot gases and of the fabric are lower, and condensation of steam and pyrolysis production occurs.

Finally, the M.I.T. group has shown that a simple linear relationship exists between the total thermal dosage received by the skin and the depth of resulting skin damage. This linear relationship holds for thermal dosages above 4.0 cal/cm^2, and it is similar to the correlation reported by the Gillette Research Institute. Below 4.0 cal/cm^2 and down to 2.0 cal/cm^2, the depth of skin damage is constant at about 0.10 mm, corresponding to a second degree burn, while no irreversible damage was observed below a thermal dosage of 2.0 cal/cm^2.

In conclusion, there is general agreement on the prediction of depth of skin damage among several groups who have recently been studying burn injury from flaming fabrics, - including Chouinard [34], GRI [R28] and M.I.T. [R25]. At the heat flux levels generated by burning textiles, initial second degree burns are to be expected in human skin which receives a thermal dosage of 2 cal/cm^2 while dosages exceeding 4 or 5 cal/cm^2 will result in linearly varying depths of dermal injury up to a dosage of 12 to 15 cal/cm^2, corresponding to a depth of damage of 2 mm.

The combustion behavior of <u>fabric</u> <u>combinations</u>, either as fabric over fabric in a specimen burn test or as double garment in a mannequin burn test has been studied by Chouinard, GRI and MIT. Chouinard noted that when several layers of cotton garments were combined in a burn test the heat dosage transferred was equivalent to that of a single cotton dress of the same total weight as the combination. However, when a nylon slip was worn beneath a cotton dress the thermal dosage transferred was not significantly different than that transferred by the cotton garment alone.

Tests of double layers of fabric spaced 1/4" from a sensor board also showed that heat transfer from double layers of cotton or PE/C fabrics was approximately the sum of transfers from the individual fabrics tested alone. The same was true when a thermoplastic fabric was placed <u>above</u> a cotton fabric during 45° upward burn tests (i.e. heat transfer from the combination burn tests was approximately the sum of the transfers from the individual fabrics tested alone). However, when the thermoplastic was placed <u>beneath</u> the cotton, the heat transfer of the combination was only slightly greater than that observed for the cotton alone. In the former case, the thermoplastic melted onto the cotton fabric and burned in place, but when the thermoplastic was underneath, it was melted by the faster burning cotton, and formed droplets which fell to the sensor surface without burning completely.

The GRI tests of combination garments with a lightweight polyester/cotton dress worn over a lightweight nylon tricot slip showed that the heat dosage transferred was about the same as for the PE/c dress tested alone. However, as heavier PE/c or cotton dresses were combined with the nylon slip, the heat dosage approached that expected from the sum of separate burnings, i.e. dependent on the combined weight.

The GRI garment tests included burning of single dresses of thermoplastics as well as dresses of fire resistant materials such as NOMEX and FR-treated cotton. These dresses did not sustain combustion when tested alone, even when ignition was attempted with a large paper fuse. When a cotton garment was combined with the flame resistant materials, the results differed with the relative positions of the two. When the FR garment was on the inside, the average heat transferred was less than for the cotton dress alone by as much as 50% (i.e. the FR garment tended to shield the skin, not enough to avoid second degree burns, but enough to reduce the area burned). When the FR garment was on the outside, the average heat transferred was greater than for the cotton garment alone, but the area of second degree burns was slightly reduced.

The design and fit of garments was shown in the GRI tests to affect the rate of flame spread and the level of heat transfer. Properly fitted slacks did not allow flames to propagate as rapidly as did dresses, while oversized slacks supported by suspenders increased the rate of flame spread as compared to dresses. Finally, belted dresses consistently transferred less heat during burning tests than did corresponding dresses of A-line style.

CHAPTER VI

SUMMARY, CONCLUSIONS AND RECOMMENDATIONS

6.1 SUMMARY   339
     General   340
     Ignition   342
     Flame Propagation   345
     Burn Injury   351

6.2 CONCLUSIONS   356

6.3 RECOMMENDATIONS FOR FUTURE WORK   362

CHAPTER VI

SUMMARY, CONCLUSIONS AND RECOMMENDATIONS

6.1 SUMMARY

We have attempted to provide a summary or overview of
each GIRCFF program phase with its corresponding chapter.
Therefore, there is no point in collecting these summary sections and repeating them here. Still, we feel that a consol-
idation of the total program results is warranted and that use
of a new format would help in its presentation.

The question and answer format appeals to us in its "small
bite" treatment of the considerable quantity of information
developed in the GIRCFF program. Clearly, it risks oversimpli-
fication, for with emphasis on relatively brief (and mostly
independent) answers, clarifying remarks and underlying assumptions and limitations of the answers are restricted and may be
omitted altogether. An effort has been made to furnish the
necessary caveats where answers appear to have possible economic or political impact. In case of doubt, the full text should
be consulted, and even the original reports listed in the reference section.

The questions were formulated on a simple criterion, i.e.,
what would the technologist in industry, in government or consumer groups, and the medical profession as well, want to know
about flammability of apparel textiles? What rules of thumb,
what rough models, what order of magnitude of results would a
technologist concerned with flammability like to have recorded
in a simple card file? What follows is our collection of 'single card items' presented in linear form and structured according to the sequence of the previous chapters. After the Q and
A summary, we present the conclusions derived from the program
including those put forward by the investigating groups and

339

those formulated by the overview team. Finally, recommendations for future work are presented.

GENERAL

1. Q. What determines the probability of burn injury to the wearer of a particular textile fabric?

   A. The overall likelihood of burn injury can be considered as a product of probabilities of exposure to ignition sources, of ignition after exposure, of flame propagation following ignition, of non-extinguishment following propagation and of injury produced by heat transfer from the burning fabric. A break in any of these probabilistic links would serve to prevent injury.

2. Q. What was the focus of the GIRCFF program on flammable fabrics?

   A. The program was concentrated on the phenomena of ignition, flame propagation and thermal injury.

3. Q. What purpose was served by focussing on the GIRCFF series of fabrics?

   A. Since four independent laboratory groups were studying different aspects of the problem, well characterized common materials were deemed essential for comparison of ignition and burn test results, and of physical and thermal measurements obtained by the several investigators.

4. Q. How were fabrics selected for the work of the GIRCFF program?

   A. It was thought desirable to investigate commercially important apparel fabrics over a broad range of fiber compositions, weights and constructions, and to include some systematic comparisons or pairs whenever possible. The selection also included both thermoplastic and non-thermoplastic materials.

5. Q. Were flame-resistant fabrics included?

A. The only commercially available flame resistant apparel fabric included at the time the program began was FR cotton flannel. This was one of the twenty fabrics selected, with an untreated cotton flannel counterpart. Another flame-resistant fabric (Nomex ^{R}) was added much later in the program and used for specific experiments.

6. Q. Are the fabrics selected in the original GIRCFF group still commercially important today?

A. Many of the fabrics used are identical with or similar to materials which are presently of commercial importance. However, flame-resistant fabrics that have become available more recently such as modacrylics, should be added for any future work.

7. Q. What kinds of physical characterizations and thermal measurements were made of the fabrics in the GIRCFF series besides ignition and burning tests?

A. In addition to the usual technological characterizations such as weave structure, fiber content, weight, permeability and yarn density, a number of thermal and optical properties were measured for the 20 test fabrics. These included specific heat, thermal conductance, ignition and melting temperatures, reflectance, transmittance and absorptance.

341

8. Q. What are the processes involved in the ignition of textiles?

   A. The main processes are:
      * heat transfer from the ignition source to the fabric,
      * thermal decomposition of the fiber polymer,
      * diffusion and convection of the products of thermal
        decomposition,
      * kinetic reactions involving decomposition products
        and oxygen from the environment.

   For thermoplastic fabrics, additional processes such as
   shrinking, melting and dripping are also involved.

9. Q. What is the effect of the thermoplastic behavior of synthetic fabrics on ignition?

   A. On heating, synthetic fabrics may shrink, melt and/or
      drip. Thus, the fabric moves away from the flame. The
      extent of these phenomena in the fabric is determined by
      its weight, structure, thermophysical properties, pre-
      drawing conditioning, orientation in the gravity field,
      by mechanical constraint (e.g. skin or another material in
      proximity), and by the heating intensity. Depending on
      how fast and how far the fabric retracts from the flame,
      it may or may not ignite. (Flame spreading may also be
      affected in a similar way.) Thus, under specific condi-
      tions, some thermoplastic fabrics can fail to ignite or
      appear self-extinguishing.

10. Q. What are common ignition sources in fabric apparel fires?

    A. Common ignition sources may be radiative heat sources
       such as electric heaters and hot plates, or flaming heat
       sources such as gas ranges, gas space heaters, matches
       and lighters.

11.Q. Why distinguish between these two types of ignition
sources?

A. Because the predominant heat transfer mechanism is
different for each case: Radiation predominates for
electric heaters and hot plates, while convection
predominates for flaming sources. Furthermore, if the
fabric is in actual contact with an electric heater or
a hot plate, solid-phase conduction becomes very important.
In real life fabric ignition accidents, heat transfer may
occur by more than one mechanism simultaneously.

12.Q. Why distinguish between heat transfer mechanisms?

A. Not all the heat given off by the ignition source is
absorbed by the fabric. The heat transfer mechanism
determines the fraction of this heat which will be
absorbed. For the cases of fabric ignition studied,
the fraction absorbed under convective heating (0.89 to
0.94) is larger than that under radiative heating
(0.18 to 0.72).

13.Q. What was the range of heating intensities used in the
ignition experiments of the GIRCFF program?

A. For radiative heating, this range was from 0.25 to 16
Watts/cm^2 (800 to 50,000 BTU/hr ft^2). For convective
heating, the heat flux is more difficult to measure and
depends greatly on the fabric-flame configuration.

14.Q. How fast does fabric ignition occur?

A. For the above heating intensities, the range of
ignition times is from a fraction of a second to 100
or more seconds.

15.Q. Can skin pain be used as a warning against fabric
ignition?

A. No, because ignition occurs before the fabric wearer
can feel any pain. The ratio of time to feel pain/time to

343

ignite depends on heating conditions. Typical ratios
observed were from 1.2 to 2.8.

16.Q. Can the time to ignite be predicted from knowledge of
material properties, heating intensity and fabric-
ignition source configuration?

A. No, the present theoretical understanding of the ignition phenomena is too crude to make such a prediction.
However, a fair estimate of a lower bound on the ignition time can be determined. A lower bound is on the
conservative side and, thus, valuable for safety considerations. This estimate is given by the following ratio:

$$\frac{\text{amount of energy needed to reach an ignition temperature}}{\text{rate of heat absorption by the fabric}}$$

This lower bound can be used to rank fabric with respect
to ease of ignition (or melting)--at any specific heating
intensity.

17.Q. What determines the amount of energy needed to reach an
"ignition temperature"?

A. This is determined by the product of weight per unit
area, heat capacity and ignition (or melting) temperature.
For the GIRCFF fabrics, weight varies from 1.90 to 8.66
oz/yd^2; the specific heat from 0.91 to 2.42 Ws/g°C; and
the ignition (or melting) temperature from 220° to 500°C.
Thus, the effect of variation in weight is predominant.

18.Q. What determines the rate of heat absorption by the fabric?

A. This depends on the heat transfer mechanism:
* for radiative heating, the rate depends on the heating intensity of the radiative source and the absorptivity of the fabric;

344

* for convective heating, (flaming ignition sources), it
  depends on the excess of the flame temperature over
  fabric temperature during the exposure interval and on
  the fluid mechanical interactions between the fabric
  and the flames.

19.Q. How do the results of bench tests compare to those of
  more realistic "sleeve ignition" results?

A. There is qualitative agreement between the results of
  these two kinds of ignition tests. Both tests show comparable effects of fiber content, fabric weight $(\frac{oz}{yd^2})$ and
  fabric-flame spacing.

FLAME PROPAGATION

20.Q. To what extent do textile fabrics exhibit different burning characteristics when exposed to an ignition source?

A. Some ignite and burn freely upon removal of the ignition
  source, some appear to ignite during exposure to the igniting source but do not sustain the flame and some appear to be unaffected by brief exposure to an ignition
  source. In addition, some fabrics shrink away from the
  ignition source, and thus avoid ignition, or if forced
  to ignite, soon melt and drip down segments of molten,
  often burning polymer. If the burning portions break
  away, the fabric may thus self-extinguish.

21.Q. What burning characteristics do undyed, untreated cotton
  fabrics exhibit?

A. The undyed untreated GIRCFF cotton fabrics sustain ignition and propagate via a uniform flame, sometimes producing a whitish smoke, sometimes relatively little visible
  smoke.

345

22.Q. What was observed for the GIRCFF flame-resistant treated cotton fabric?

A. It did not propagate a flame when tested alone, nor did the aramid, NOMEX fabric tested.

23.Q. Do the polyester/cotton blends exhibit burning characteristics similar to cotton or to the thermoplastics?

A. PE/c blends in the GIRCFF series propagate flame at velocities comparable to those of all cotton fabrics. However, they may generate considerable smoke as for certain of the thermoplastics. Furthermore, depending on blend composition, they may melt ahead of the flame and, in some cases, actually drop molten burning polymer.

24.Q. What is the flame propagation behavior of thermoplastic fabrics?

A. It depends on the fiber composition, the fabric structure and the proximity of a non-thermoplastic fabric. When tested alone, certain lightweight thermoplastic-fiber fabrics will not sustain combustion, however certain heavyweight thermoplastics burn, with considerable black smoke.

25.Q. Which of the GIRCFF thermoplastic fabrics do not support combustion when tested alone?

A. In laboratory specimen tests, the lightweight 100% nylons and 100% polyesters shrink from the igniting flame and melt, so that burning is not sustained. In garment tests, the lightweight nylon dresses and polyester dresses worn by themselves do not sustain combustion even when ignition is attempted with a large paper fuse.

26.Q. What about the other thermoplastics in the GIRCFF series? How do they burn?

A. The heavy double-knit polyester propagates an irregular

narrow flame leaving most of the specimen width unburned.
The Acetate propagates a flame in a zig-zag pattern. The
acrylic burns with a characteristic odor, with black
smoke and a high flame, and with continued burning of
fallen pieces or melted drops. The acetate/nylon blend
burns irregularly leaving islands of unburnt fabric.

27.Q. How does wool perform in burning tests?
   A. The one wool fabric tested does not support combustion in
      a horizontal test, but in a 45° upward test it burns in
      a narrow zone leaving most of the specimen width unburnt.

28.Q. What construction factors influence the rate of flame
      propagation in 100% cotton fabrics?
   A. Fabric weight (per unit area) has a dominant effect on
      flame propagation velocity in cotton fabrics: the hea-
      vier the fabric, the slower the flame propagation. In
      horizontal burn tests with 1/2" spacing between fabric
      and a skin simulant the velocities ranged from 20 cm/min
      for a 2 oz cotton batiste to about 6 cm/min for a 9 oz
      denim.

29.Q. What other structural features have an effect on flame
      speed in burning fabrics?
   A. The porosity of fabrics (i.e. their relative air to fiber
      volume) influences flame velocities, but only at the ex-
      treme high range of porosity. None of the GIRCFF test
      fabrics fall in this range.

30.Q. What conditions of test burning in air influence flame
      velocities in cotton fabrics?
   A. The principal test conditions which affect flame veloci-
      ties are sample orientation and direction of burning, and
      spacing between fabric and skin simulant. In horizontal

347

1/2" spaced burning of cotton fabrics in the weight
range investigated, (2 to 9 oz/yd^2) the flame propagates
from 20 to about 6 cm/min. For burning at 45° upward,
the velocities of propagation range from about 300 to
about 60 cm/min--a tenfold increase. For 90° upward
burning of the GIRCFF cotton fabrics, flame propagation
velocity is similar to that observed for the 45° upward
test.

31.Q. How do polyester-cotton blends compare with the all-
cotton fabrics in regard to flame velocities?

A. The PE/c blends show a similar effect of fabric weight
on flame propagation velocities. The PE/c blends
propagate flame a few cm/min faster than cotton fabrics
of comparable weight in horizontal burning. The blends
burn somewhat slower in the 45° upward burn test and
somewhat faster in the 90° upward burn test.

32.Q. How does fabric/skin spacing affect flame speeds?

A. For the 90° upward burning of a 2 oz/yd^2, 100% cotton
fabric, the flame velocity increases linearly from
about 80 to 280 cm/min as the spacing is increased from
1/8" to 1/2"; for a 4 oz/yd^2 fabric, it increases
linearly from about 20 to 140 cm/min.

33.Q. How do those fabrics which did not support flame when
tested alone, perform burn tests with other fabrics
which do, of themselves, propagate flame?

A. They burn, but at a slower rate. This is expected
because of the higher combined weight. When a series of
GIRCFF thermoplastic fabrics is matched up with an all-
cotton knit T-shirt material and burned 45° upward at
1/2" spacing to the skin simulant (with the cotton underneath), the velocity of flame propagation ranges from 50%
to 85% of that measured for the cotton alone, depending
on the weight of the thermoplastic. The burning proceeds

348

stepwise, reflecting a cyclic interaction between the
two fabrics with the thermoplastic first shrinking,
then melting, then igniting in built up areas and continuing to burn until fully consumed.

34.Q. What about wool in combination with a cotton T-shirt
material?

A. In the single case observed, it burned at 1/2 the
velocity of the cotton alone, as one would expect from
its heavier weight. It also gave off the most heat.

35.Q. To what extent do garment-mannequin tests parallel
laboratory specimen burn tests?

A. Single garment burn tests with the GIRCFF cotton fabrics
and polyester/cotton blend fabrics manifest extensive
flame propagation, resulting in burn-causing heat-
transfer. As in the case of laboratory specimens,
weight is a dominant factor in slowing down flame
propagation rates. Garment fit which corresponds to
fabric-skin spacing is also a significant factor in
garment-mannequin tests.

36.Q. How do all cotton single garments compare with polyester/
cotton garments?

A. The cotton is observed to burn more quickly than the
PE/c blends, but both materials burn to the same
ultimate area for fabrics of comparable weight. It
should be stressed that direct measurements of flame
propagation were not made in the mannequin tests. The
above conclusion is based on the rate of increase of
burn-area in the mannequin body.

37.Q. What features of garment design affect flame velocities
in mannequin burn tests?

A. Primarily, tightness of fit affects flame speed whether
by garment design (as in a belted vs. unbelted dress) or
in size selection. The looser the garment, the faster

349

the flame spread--corresponding to the linear relation
between fabric-skin spacing and flame velocity in the
laboratory tests.

38.Q. How do the thermoplastic fabric dresses propagate flame
in mannequin tests?

A. In the GIRCFF fabric series, the lightweight nylon and
polyester garment tested by itself did not sustain
combustion even when ignition was forced.

39.Q. How about the flame resistant fabrics tested in a
garment burn experiment?

A. The flame resistant-treated cotton of the GIRCFF
series, and the supplemental sample of NOMEX did not
sustain combustion.

40.Q. What if garments made from thermoplastic fabrics are
tested in combination with a cotton or a PE/c undergarment?

A. When cotton or PE/c blend fabrics are used for an outer
garment or dress, and nylon for a slip, the extent of
flame propagation in 30 secs is comparable to that of
a single cotton or PE/c garment of weight equivalent
to that of the combined garments. A similar result
was obtained for the one test combination consisting of
a polyester dress and a polyester/cotton slip. In
laboratory tests of combinations of thermoplastic
fabrics placed on top of a cotton underfabric, the 45°
upward burning rates were in general agreement with
the flame velocity for equivalent weights of
all-polyester/cotton fabrics.

41.Q. How do flame-resistant fabrics behave when burned in
combination with cellulosics or PE/c fabrics?

A. When the cellulosic is used for the outer garment and
the flame-resistant fabric for the inner slip, the area
burned in the test is only slightly less than for a
burn test of the cellulosic garment alone. When the

flame-resistant fabric is used in the outer garment and the cellulosic in the slip, the area of burn is reduced still further. The effect is similar in test combinations involving either flame resistant cotton or NOMEX in combination with the cellulosic.

BURN INJURY

42.Q. How much heat does an ordinary flammable textile fabric generate when burned?

A. When burned vertically in air at 21% O_2 concentration, fabrics will give off from 2000 to over 4000 cal/g. For cotton, the value is of the order of 3000 cal/g.

43.Q. How does heat of burning calculate out for commercial weight fabrics (such as those included in the GIRCFF series) in terms of cal/cm^2?

A. For fabrics weighing from 2 to 10 oz/yd^2 the heat of burning in air ranges from 20 to 100 cal/cm^2.

44.Q. What portion of this heat is developed and reaches the skin when fabric specimen or clothing is ignited?

A. It depends on the conditions of burning, e.g. the orientation, position, and spacing of the fabric vs. the skin or skin simulant, and the fiber type: Laboratory tests in 45° upward burning with a skin simulant set 1/4" below show that 10 to 15% of the heat generated reaches the skin. In vertical tests with 1/8" to 1/2" spacings, the percentages run from 15 to 35% or, in absolute terms, 5 to 15 cal/cm^2. In the case of garment tests on mannequins, the average heat dosage which reaches a majority of sensors embedded in the torso ranges from 25 to 50% of the heat of burning (for the cotton and PE/c fabrics of the GIRCFF series). This corresponds to heat dosages of 10 to 25 cal/cm^2.

351

45.Q. Why the difference in heat dosages from the same fabrics in laboratory and garment tests?

A. The results of the garment tests as reported, are based on average heat dosages received by those sensors which experience more than the threshold of 2 cal/cm^2, thus weighting readings in the hotter areas during the garment burn. Furthermore, the garment burn differs from laboratory specimen vs. sensor board tests with respect to length of free hanging material, spacing to the skin, changing directions of fabric orientation, and presence of chimney effects.

46.Q. What heat dosage will cause skin injury?

A. A rough estimate of critical dosage for second degree burns[*] is 2 cal/cm^2 at heat fluxes between .25 to .75 cal/cm^2 sec. Thus, all of the fabrics in the GIRCFF series would cause second degree burns, or worse, on human skin if and when they were to propagate a flame in garment form.

47.Q. What if the heat dosage exceeds 2 cal/cm^2?

A. From 2 to 4 cal/cm^2 still causes second degree burns[*], but for 4 cal/cm^2, and above, one suffers the beginning of third degree burns as tissue destruction penetrates into the dermal component of the skin. Tissue destruction is roughly linear with thermal dosage down to the base of the dermal layer.

---

[*]According to medical convention, a first degree burn involves the epidermis, and theoretically could involve the entire epidermis. A second degree burn involves the entire epidermis, plus some portion of the dermis, and these have been spoken of as superficial second degree burns, or deep second degree burns, depending on the depth of invasion of the dermis. The third degree burn destroys the entire dermis and epidermis, so that the destruction then involves the subcutaneous tissue [54].

48.Q. Is heat dosage of itself the prime determinant of skin
      burn injury?
   A. Heat dosage is only a rough determinant of skin damage.
      Heat must be absorbed by the skin at a rate above a
      threshold value in order to cause damage. In fact the
      human skin can withstand infinite thermal dosages at
      extremely low flux levels.

49.Q. What is the nature of thermal injury in the skin?
   A. Tissue death or necrosis is observed above a threshold
      temperature of 44°C. Damage occurs as a first order rate
      process and is dependent on the temperature above 44°C.*
      Total injury at a given tissue depth can be expressed as
      a damage integral based on the time temperature history
      at that level.

50.Q. What is the nature of heat transfer from a burning fabric
      to the skin?
   A. The total flux to the skin is composed of radiative, gas
      conductive, convective and condensative components. In
      addition when melting and dripping of thermoplastic fabrics
      occurs heat transfer may also take place by direct contact
      conduction between the polymer and the skin.

51.Q. What proportion does each component contribute?
   A. It depends on the conditions of burning i.e. whether hori-
      zontal or vertical, what spacing is left between fabric
      and skin, and the point in time of burning. For example,
      in vertical burn tests of cotton fabric spaced 1/2 inch
      from the skin at the time of maximum heat flux, radiation
      and gas conduction each account for about 50% of the heat
      transferred. For 1/8 inch spacing, radiation, conduction
      and condensation can each contribute about 33%.

---

*Lower temperatures, however, cause irreversible injury to cells
and/or generate inflammatory reactions which in themselves des-
troy tissue, so that tissue destruction over time can be accom-
plished below 44°C [54].

52.Q. What temperatures are generated when a fabric burns in proximity of the skin?

A. For cotton fabrics burned upward vertically 1/2" from a skin simulant, maximum air gap temperatures are about 1200°C. As the gap is reduced to 1/8", the maximum gap temperature is reduced to 500-600°C and no flame is visible in the air gap.

53.Q. Does one need to use animal skin to measure burn injury resulting from burning fabrics?

A. Only for the initial purpose of correlating damage integrals and depth of skin damage as determined by biopsic sections.

54.Q. What means can be used to meter heat transfer leading to skin burn injury?

A. A skin simulant in the vicinity of the burning fabric may be used to "stretch" distances and thus facilitate measurement of equivalent deep tissue temperature-time-histories for use in the damage integral. A simpler method is to measure heat flux vs time at a surface sensor in the simulant surface.

55.Q. How is the time history of heat flux received at the skin surface convertible to injury depth?

A. By the direct rough correlation mentioned earlier. A more sophisticated method has been worked out whereby the surface flux history is combined with the optical and thermal properties of the epidermal and dermal layers (including the effect of blood flow) to compute the time-temperature histories at various tissue depths.

354

56.Q. How do the GIRCFF fabrics perform with respect to heat
generated when subjected to burning in combinations?

A. The heat generated in burning of two layers of cotton
fabric or two layers of PE/C fabric is comparable to that
from burning a single fabric of weight equal to the sum
of the two individual fabrics. The flame speed is reduced, and the heat dosage transferred is increased as
for a single heavy weight fabric. The duration of burn
and the average heat transferred increase linearly with
the weight of the burning fabric, whether in single or
multiple layers.

57.Q. How do the lightweight thermoplastic fabrics perform when
combined with a cotton or PE/C?

A. If the thermoplastic is used for the inner garment and
cotton for the outer garment, the average heat transferred
is only slightly greater than for the cotton garment alone.
If the thermoplastic is used for the outer garment, the
heat transferred corresponds closely to the sum of the
heats transferred in single burn tests of the two fabrics.

58.Q. How does FR cotton or Nomex behave in composite garment
tests?

A. If the flame resistant garment is worn inside and a cotton garment outside, the heat transferred and the area
burned are less than for the cotton garment alone, indicating a shielding effect by the fire resistant slip.
If the fire resistant garment is worn outside, the average heat transferred is slightly higher than for the cotton alone, but the area of burn is less. Furthermore,
when the flame resistant garment is worn outside the hazard of ignition and consequent flame propagation is less.

6.2 CONCLUSIONS

There are numerous links in the probablistic chain of
events between the setting of textile standards for flammability and the occurrence of burn injuries in a population of
garment wearers. The GIRCFF program has focussed on two elements in the series of events which lead to human injury from
burning clothing: (a) the probability of ignition of a given
fabric or garment given a specific thermal exposure and, (b)
the probability of burn injury to the wearer of the garment
following fabric ignition.

Flame propagation, a necessary condition preceding significant burn injury, is implied either in element (a) ignition
or element (b) burn injury. Trade standards and government
specifications are generally concerned with the former, i.e.,
with determining in single tests whether a fabric subjected to
a given heat exposure will ignite and propagate a flame. In
contrast ignition has been studied separately in the GIRCFF
program, - as a transient phenomenon preceding continuing flame
propagation. In another phase of the GIRCFF program, heat
transfer from steadily burning fabrics has been studied as the
cause of burn injuries.

The findings of the latter program phase indicate that
when flame propagates in garments constructed of commercial
weight fabrics, the skin of the wearer will receive sufficient
heat dosage   to cause second degree burns or worse. The area
of skin thus injured will correspond closely to the area of
fabric burned. These results are in agreement with data reported by other investigators, and suggest that test standards and
material characterization which quantitify the ignition and
flame propagation behavior of apparel fabrics are sufficient of
themselves to indicate burn injury potential. Heat-transfer
quantification is of lesser importance. Ease of ignition and
flame propagation in fabrics of commercial weight are therefore
the key factors in burn hazard to the apparel user.

356

The finding of the studies on ignition of a series of commercially important fabrics indicates that virtually all of the
GIRCFF materials tested will ignite under some conditions, when
subjected to an open flame. And once ignited, virtually all of
these fabrics will, under some conditions, propagate flame over
a major portion of their surface area, thus causing potential
burn injury to any adjacent human body. This is admittedly a
negative statement of the program results but it is a useful
caveat for both producer and user of currently commercial textile materials.

On the other hand, a more optimistic view of the GIRCFF results can be taken, - namely that some commercial fabrics will
not ignite under certain conditions and thus will not threaten
burn injury to the garment wearer. Or that some fabrics under
certain conditions of usage and exposure may ignite, but will in
a short time self extinguish, thus significantly reducing burn
hazard. Stated another way, the GIRCFF results stress the fact
that many commercial materials which may appear to be safe under
one condition of use and exposure, may be hazardous under different circumstances.

The original concept of determining the probabilities of
igniting a given fabric subjected to specified thermal exposures
and the probability of its propagating flame is still valid.
The ignition step may be viewed in an oversimplified way as a
transfer of sufficient heat to the fabric as to raise its temperature to a level conducive to ignition. If the physics of
the situation is thus established, even in such a crude model,
then the uncertainty of igniting should be related to the usual
variability of structural and thermal and kinetic properties of
commercial fabrics which determine the magnitude of heat dosage
necessary for causing the requisite temperature rise for ignition of a particular specimen of fabric. Or the uncertainty of
ignition may relate the actual variability in heat flux reaching
the fabric, compared to the specified level of heat flux.

357

The GIRCFF program did determine the frequency of
ignition of certain test fabrics under a few select conditions
of exposure. But the major task of determining extensive
ignition probabilities vs. exposure conditions for a range of
commercial fabrics has not been undertaken. Instead, the GIRCFF
program has identified for its test series the average level
of heat dosage necessary for fabric ignition under exposure to
known conditions of heat transfer. Further, the GIRCFF
investigators have identified the thermal conditions and the
specific physical circumstances under which the common commercial
fabrics do or do not ignite and propagate flame. Thus, while
use of flame resistant materials offers the most secure route
to garment fire safety, these results offer some promise for
alleviating the fire hazard present in untreated, commercially
available materials.

For example, the GIRCFF results indicate that when a
fabric is exposed to a high-level heat flux for a fixed period
of time, ignition can be avoided if the heat required for fabric
ignition is, by design, increased above that amount of heat
which is actually delivered to and absorbed by the fabric. The
situation can be attained in the case by selecting heavier
fabrics and thus increasing the product of mass and specific
heat or increasing the sensible heat needed for ignition (as
illustrated in the effect of cotton fabric or PE/c fabric
weight on ignition behavior). Or the likelihood of ignition can
be lowered by reducing the amount of heat delivered to and
absorbed by the fabric (as occurs when a thermoplastic fabric
shrinks and melts during exposure to a heat source and thus
mechanically reduces the amount of heat transferred).

Either of these cited examples represents a valid method
of reducing ignition hazard. The same two phenomena also
affect the mechanism of flame propagation, since flame propagation
requires continuous re-ignition of the fabric ahead of the flame
front. Added combustible mass slows down the temperature increase

358

and therefore the ignition of new portions of the sample. Thus heavier weight fabrics propagate flame more slowly. And if fabric ahead of the flame melts extensively, the burning portion of the sample may, if unsupported mechanically, break away, thus removing the flame from the remainder of the specimen. A flame extinguished by any means is still out.

The objective of increased safety will not be served however, unless it is emphasized that added fabric mass will not help in the ignition problem if time of exposure to a given heat source is increased indefinitely(case of the child climbing near the electric heater). And the shrinking and melting characteristic will lose a good deal of its effectiveness if shrinkage is restrained by too tight a fabric construction or if a non-melting mechanical support is provided for the heated thermoplastic fabric (as in the case of using nonthermoplastic stitching, or in the presence of an adjacent cellulosic fabric).

Finally it should be noted that flame resistant materials such as FR treated cotton can be ignited, although they have little tendency to propagate flame when tested alone. Under these 'pure' conditions the FR treated material affords maximum protection from garment burning injury. When worn under mixed conditions (as with a cellulosic slip) the GIRCFF FR treated fabric did burn and gave off enough heat to cause severe burns over large areas of the body. Nevertheless the burn area and the heat dosage observed in tests of FR garments worn over or under cellulosic fabrics are well below that for all cellulosic untreated fabrics of equivalent weight or for mixtures of cellulosic and thermoplastic fabrics of equivalent weight.

The more specific conclusions relating to the <u>ignition phase</u> of the GIRCFF study are summarized as follows:

* Ignition in medium weight fabrics occurs well before the wearer feels pain in thermal exposure to radiative or flaming heat sources. Thus, a heat activated warning pain can not be expected to trigger evasive action by the wearer before his clothing is on fire.

    * Ignition times of untreated fabrics depend directly
on variables such as fabric weight, and fiber content.
They are likewise influenced by exposure conditions such
as heat source characteristics, and position relative to
the fabric. Also the presence of a skin simulant and of
other fabric layers has some effect.
    * A semi-empirical correlation has permitted the devel-
opment of expressions to predict ignition or melting
times, expressions formulated in terms of fabric thermal
and physical properties and conditions of thermal exposure. Such expressions are of value in providing direction for improved design of fabrics and modification of
fibrous materials. Variations of the physical models on
which the expressions are based permit the treatment of
either radiative or of convective (flaming) heating modes.

Conclusions relating to the flame propagation phase of the
GIRCFF studies are summarized below:
    * Velocity of flame spread in cellulosic fabrics is in-
versely related to overall weight of the single fabric
or fabric combination involved in a burn test. The same
is true for a combination of untreated cellulosics and
thermoplastics where the cellulosic interferes with the
self extinguishing mechanism observed in single lightweight thermoplastics.
    * Velocity of flame spread in fabrics burning in proxi-
mity of a skin simulant increases with increased spacing
between fabric and skin. Orientation in the gravity
field can account for a ten fold increase in flame velocity, horizontal burning proceeding more slowly than upward burning.

Conclusions relating to heat transfer from burning fabrics
are summarized below:
    * The total heat dosage received by a skin simulant

360

placed near a burning fabric is highly dependent on the circumstances of burning, such as the specimen orientation and the fabric-skin spacing. If these conditions are kept constant, the rate of heat transfer from a given fiber composition appears to be affected only slightly by the weight and construction of the fabric being burned. Thus total heat dosage is primarily a function of the overall time of thermal exposure, in other words, a function of the flame velocity and therefore of fabric weight.

* The state of the art has been well advanced for instrumentation of fabric specimen and garment burn tests. Mannequins with distributed sensors are available and permit measurement of heat dosage and equivalent burn areas resulting from static burning garments. Sophisticated skin simulants are available with instrumentation for measurement of heat fluxes and total dosages coming from adjacent burning fabric specimens, and also to determine temperature histories at subsurface skin locations.

* The application of skin simulant thermal data to composite mathematical models of the skin permits the computational conversion of heat flux data to predictions of depths of real skin burn injury which will result from such exposure. Above a given threshold level, depths of burn injury computationally derived and also biophysically determined are shown to be linearly related to total heat dosage. Thus total heat dosage can be taken as a rough measure of thermal skin injury.

6.3 RECOMMENDATIONS FOR FUTURE WORK

Our recommendations are prepared in two categories: (a) for future work to expand and complete certain unfinished work phases of the present program and (b) for initiation of new areas of research intended to minimize burn injury potential for wearers of apparel fabrics. For continuity with the summary material we consider the unfinished work area first.

We recommend that future programs on fabric flammability undertaken under joint industry and government sponsorship include:

* Further development and validification of an ignition parameter or index for textile materials based on fabric thermal and physical properties. Such an index could lead to quantification of the heat dosage needed to ignite textile materials and thus could be a useful guide in future product development.

* Given the nature of structural changes which occur during heating of thermoplastic fabrics (and to some degree non thermoplastics), future measurements of physical, as well as thermal fabric properties should be conducted over a range of temperatures up to and including the combustion point.

* Further development of the reaction kinetics of pyrolysis and oxidation of specific materials as affecting ignition.

* Develop improved laboratory tests capable of measuring ignition and flame spreading behavior of thermoplastic as well as non-thermoplastic materials.

* Further study the effect of fabric structure on the ignition and burning of thermoplastic fabrics and garments.

* Extend the study of ignition and burning hazards in
flame-resistant textiles to include fibers, finishes and
fabrics representative of current commercial developments.

* Further study the ignition and burning behavior of
fabric combinations which include 'solo-burning' fabrics
and 'solo-self-estinguishing' fabrics; e.g., combinations
of cellulosics with thermoplastics or flame resistant
fabrics.

In study areas not covered by the original GIRCFF projects,
we recommend that future industry-government programs include
the following phases:

* Determine the statistical nature of thermal exposure
experienced by various segments of the population in normal living and working situations.

* Quantify the nature of thermal exposure experienced
by persons involved in accident situations involving fire;
e.g.,automobile crashes, home fires, etc.

* Determine the extent to which statistical variations
in fabric properties around given nominal values influence ignition times and flame velocities.

* Determine the effect of air movement and of dynamic
changes in mannequin configuration during simulated garment fires.

* Study the feasibility of laboratory tests of a type
between single fabric strip tests and full-scale mannequin garment burn tests. Such methods should include
consideration of the effects of skin simulants in proximity to the burning fabric.

* Study the effect of varying shrinkage and melting
characteristics of thermoplastic fibers on ignition and
flame propagation in fabric and garments.

* Study the effect of contaminants such as those generated in use and in servicing upon ignition and burning
of thermoplastic fibers.

* Study the factors which influence ease of mechanical
extinguishment of burning clothing.

GIRCFF REPORT REFERENCES[*]

R1.   Alkidas, A., Hess, R.W., Kirkpatrick, C.S.,
      Kirkpatrick, M.J., Bergles, A.E., Wulff, W. and Zuber, N.,
      "Study of Hazards from Burning Apparel and the Relation
      of Hazards to Test Methods", Progress Report No. 1,
      Georgia Institute of Technology, April 1, 1971.

R2.   Alkidas, A., Hess, R.W., Kirkpatrick, C.S., Bergles, A.E.,
      Wulff, W., and Zuber, N., "Study of Hazards from Burning
      Apparel and the Relation of Hazards to Test Methods",
      Progress Report No. 2, Georgia Institute of Technology,
      July 1, 1971.

R3.   Alkidas, A., Hess, R.W., Kirkpatrick, C.S., Mays, R.L.,
      Wulff, W. and Zuber, N., "Study of Hazards from Burning
      Apparel and the Relation of Hazards to Test Methods",
      Progress Report No. 3, Georgia Institute of Technology,
      October 1, 1971.

R4.   Alkidas, A., et al, "Study of Hazards from Burning
      Apparel and the Relation of Hazards to Test Methods",
      Final Report, Georgia Institute of Technology,
      December 1971, (NTIS:COM-73-10954).

R5.   Alkidas, A., et al, "Study of Hazards from Burning
      Apparel and the Relation of Hazards to Test Methods",
      Progress Report No. 4, Georgia Institute of Technology,
      April 1, 1972.

R6.   Alkidas, A., et al, "Study of Hazards from Burning
      Apparel and the Relation of Hazards to Test Methods",
      Progress Report No. 5, Georgia Institute of Technology,
      July 1, 1972.

---

[*]The following initials are used in GIRCFF Report References:
FMRC for Factory Mutual Research Corporation; M.I.T. for
Massachusetts Institute of Technology; GRI for Gillette
Research Institute.

R7.   Champion, E., Giddens, W., Hess, R. W., Durbetaki, P.
and Wulff, W., "Study of Hazards from Burning Apparel and
the Relation of Hazards to Test Methods", Progress Report
No. 6, Georgia Institute of Technology, October 1, 1972.

R8.   Alkidas, A., Champion, E. R., Giddens, W. E., Hess, R. W.,
Kumar, B., Naveda, G. A. A., Durbetaki, P., Williams, P. T.,
and Wulff, W. "Study of Hazards from Burning Apparel and
the Relation of Hazards to Test Methods", Second Final Report, Georgia Institute of Technology, December 31, 1972.
(NTIS:COM-73-10956).

R9.   Kalelkar, A. S. and Kung, H. C., "A Study of Pre-Ignition
Heat Transfer Through a Fabric-Skin System Subjected to a
Heat Source", Quarterly Report 1, FMRC Serial No. 19967,
1971.

R10.   Kalelkar, A. S. and Kung, H. C., "A Study of Pre-Ignition
Heat Transfer Through a Fabric-Skin System Subjected to a
Heat Source", Quarterly Report No. 2, FMRC Serial No. 19967,
1971.

R11.   Kalelkar, A. S. and Kung, H. C., "A Study of Pre-Ignition
Heat Transfer Through a Fabric-Skin System Subjected to a
Heat Source", Quarterly Report No. 3, FMRC Serial No. 19967,
1971.

R12.   Heskestad, G., et al., "A Study of Pre-Ignition Heat Transfer Through a Fabric-Skin System Subjected to a Heat Source",
Annual Report, FMRC Serial No. 19967, 1971. (NTIS:COM-73-
10959).

R13.   Heskestad, G. and Kung, H. C., "Pain Time Versus Fabric
Ignition Time for Exposure to Flame", Technical Memorandum
FMRC Serial No. 19967, December 1971.

R14. Heskestad, G., "Ease of Ignition of Fabrics Exposed to Flaming Heat Sources", First Quarterly Progress Report, FMRC Serial No. 19967, April 1, 1972.

R15. Heskestad, G., "Ease of Ignition of Fabrics Exposed to Flaming Heat Sources", Second Quarterly Progress Report, FMRC Serial No. 19967, July 1, 1972.

R16. Heskestad, G., "Ease of Ignition of Fabrics Exposed to Flaming Heat Sources", Third Quarterly Progress Report, FMRC Serial No. 19967, October 1, 1972.

R17. Heskestad, G., "Ease of Ignition of Fabrics Exposed to Flaming Heat Sources", Final Report, FMRC Serial No. 19967, January, 1973 (NTIS:COM-73-10955).

R18. Williams, G. C., et al., "Measurement of Flammability and Burn Potential of Fabrics", Progress Report No. 1, DSR 72894, NSF Grant No. GK-27188, Fuels Research Laboratory, M.I.T., Cambridge, Massachusetts, 1971.

R19. Williams, G. C., Mehta, A. K., Wong, F., "Measurement of Flammability and Burn Potential of Fabrics", Progress Report No. 2, Project DSR 72894, NSF Grant No. GK-27188, Fuels Research Laboratory, M.I.T., Cambridge, Massachusetts, June 30, 1971.

R20. Williams, G. C., Mehta, A. K., Wong, F., Padia, A., "Measurement of Flammability and Burn Potential of Fabrics", Progress Report No. 3, Project DSR 72894, NSF Grant No. GK-27186, Fuels Research Laboratory, M.I.T., Cambridge, Massachusetts, October 1, 1971.

R21. Williams, G. C., et al., "Measurement of Flammability and

367

Burn Potential of Fabrics", Summary Report, Project DSR
72894, NSF Grant No. GK-27188, Fuels Research Laboratory,
M.I.T., Cambridge, Massachusetts, February 15, 1972.
(NTIS:COM-73-10950).

R22.   Williams, G. C., Mehta, A. K., Wong, F., "Measurement of
        Flammability and Potential Burn of Fabrics", Progress
        Report No. 5, Project DSR 73884, NSF Grant No. GI-31881,
        Fuels Research Laboratory, M.I.T., Cambridge, Massachu-
        setts, April 4, 1972.

R23.   Williams, G. C., Howard, J. B., Mehta, A. K., Wong, F.,
        "Measurement of Flammability and Potential Burn of Fabrics",
        Progress Report No. 6, Project DSR 73884, NSF Grant No.
        GI-31881, Fuels Research Laboratory, M.I.T., Cambridge,
        Massachusetts, July 5, 1972.

R24.   Williams, G. C., et al., "Measurement of Flammability and
        Potential Burn of Fabrics", Progress Report No. 7, Project
        DSR 73884, NSF Grant No. GI-31881, Fuels Research Labora-
        tory, M.I.T., Cambridge, Massachusetts, September 22, 1972.

R25.   Williams, G. C., Mehta, A. K., and Wong, F. "Measurement
        of Flammability and Burn Potential of Fabrics", Summary
        Report, DSR Project 73884, NSF Grant No. GI-31881, Fuels
        Research Laboratory, M.I.T., Cambridge, Massachusetts,
        February 15, 1973 (NTIS:COM-73-10960).

R26.   Krasny, J. F. and Fisher, A. L., "Study of Hazards from
        Burning Apparel and the Relation of Hazard to Test Methods:
        Flammability of Clothing Assemblies", First Annual Review
        Report for GIRCFF, G.R.I., January 19, 1972 (NTIS:COM-73-
        10953).

R27.   Fourt, L.,"Study of Ignition and Exposure", Final Report
       for GIRCFF, Gillette Research Institute, December 31,
       1971 (NTIS:COM-73-10958).

R28.   Arnold, G., Fisher, A., and Frohnsdorff, G., "Hazards
       from Burning Garments, Final Report for GIRCFF, Gillette
       Research Institute, March 1973 (NTIS:COM-73-10957).

REFERENCES

[1]   Ciba Review, Flammability of Fabrics 1969/4, entire issue.

[2]   Little, R.W., Flameproofing Textile Fabrics, Reinhold
       Publishing Corp., N.Y.C. (1947).

[3]   Ryan, J.V., Don't Get Burned, Polymer News, $\underline{I}$, No. 6/7,
       10/18, (1973).

[4]   Tribus, M., Decision Analysis Approach to Satisfying the
       Requirements of the Flammable Fabrics Act, ASTM Standard-
       ization News, February, 1973, p. 22-27.

[5]   Information Council on Fabric Flammability, Proceedings of
       the Third Annual Meeting, 1969.

[6]   World Textile Abstracts, Annual Index, 1972.

[7]   Lyons, J., Chemistry and Uses of Fire Retardants, Wiley-
       Interscience, N.Y.C. (1970).

[8]   Preston and Economy (Editors), Applied Polymer Symposia-
       High Temperature and Fire Resistant Fibers, APPSBX
       21-1-226 (1973).

[9]   Suchecki, S., Flammability-Let the Maker Beware, Textile
       Industries, August 1973, pages 37-63.

[10]   Hearle, J.W. W., Grosberg, P. and Backer, S., Structural
       Mechanics of Fibers Yarns and Fabrics, Wiley-Interscience,
       New York (1969).

[11]   Backer, S. and Valko, E. I., Thesaurus of Textile Terms
       Covering Fibrous Materials and Processes, the M.I.T. Press,
       Cambridge, 1969.

[12] Bernskiold, A., Ignition and Burning Properties of Textiles, Doctoral Thesis, Chalmers Tekniska Hogskola Goteborg (1970).

[13] Setchkin, N.P., A Method and Apparatus for Determining the Ignition Characteristics of Plastics, Journal of Research of the National Bureau of Standards, Research Paper RP2052, Vol. 43, 591-605, (1949).

[14] Kreith, F., Principles of Heat Transfer, International Textbook Company (1965).

[15] Carborundum Company, Brochure, Polymer Systems Department.

[16] Bates, J.J., and Monahan, T.I., Thermal Radiation Damage to Cloth as a Function of Time of Exposure, Material Laboratory, New York Naval Shipyard, June 1950.

[17] Pattern, G.A., Ignition Temperature of Plastics, Modern Plastics, p. 119, July 1961.

[18] Yeh, K.N., Report on Calorimetric Studies of GIRCFF Fabrics, Personal Communications, NBS (1973).

[19] Hearle, J.W.S. and Miles, L.W.C., The Setting of Fibers and Fabrics, Merrow Publishing Company, Watford Herts, England (1971).

[20] Valk, G., Berndt, H.J., and Heldemann, Neue Ergebnisse zur Fixierung von Polyesterfaseon, Chemiefasern Heft 5/1971, p. 1-8.

[21] Postle, R., Burton, P., and Chaikin, M., J. Textile Institute 55, T388 (1964).

[22] Arghyros, S., Heat Setting of Polyester and Nylon Monofilaments, Sc.D. Thesis, Fibers and Polymers Division, M.I.T., June 1973.

[23] Whitman, R., Textile Research Journal 17, 148 (1947).

[24] Kilzer, F. J. and Broido, A., Speculations on the Nature of Cellulose Pyrolysis, Pyrodynamics, 2, 151-163, 1965.

[25] Kanury, A. Murty, Ignition of Cellulosic Solids - A Review, Technical Report, Factory Mutual Research Corporation Serial No. 19721-7, Nov. 1971.

[26] U.S. Dept. of Health, Education and Welfare, Flammable Fabrics, Fourth Annual Report, fiscal year 1972.

[27] Carslaw, H. S. and Jaeger, J. C., Heat Conduction in Solids, Oxford Press, N.Y., 1950.

[28] Stoll, A. M., Techniques and Uses of Skin Temperature Measurements, Annals, New York Academy of Science, 121, 49, 1969.

[28a] Wulff, W. et al. Ignition of Fabrics Under Radiative Heating, Combustion Science and Technology, 1973, Vol. 6, pp. 321-334.

[28b] Wulff, W. et al, Fabric Ignition, Textile Research Journal, Vol. 43, No. 10, October 1973, pp. 577-588.

[29] Chen, N. Y., Transient Heat and Moisture Transfer to Skin through Thermally Irradiated Cloth, Ph. D. Thesis, Dept. of Chem. Engrg., M.I.T., Cambridge, Mass., Feb, 1959.

[30] Hallman, J. et al., Ignition of Polymers, Society of Plastic Engineers, J, 28, p. 43-47, September, 1972.

[30a] Wulff, W. and Durbetaki, P., Fabric Ignition and the Burn Injury Hazard, proceedings, 1973 International Seminar on Heat Transfer from Flames, Trogir, Yugoslavia, to be published by Scripta Publishing Co. (1974).

[31]   Moussa, N.A., Toong, T.Y., and Backer, S., An Experimental
       Investigation of Flame Spreading over Textile Materials,
       Combustion Science and Technology, Vol. 8, No. 4,
       p. 165, 1973.

[31a] Toong, T.Y., A Theoretical Study of Interactions between
       Two Parallel Burning Fuel Plates, Combustion and Flames 5,
       p. 221-227 (1961).

[32]   Wulff, W., Personal Communication, June 1973.

[33]   Durbetaki, P., Personal Communication, September 1973.

[34]   Chouinard, M.P., Knodel, D.C., and Arnold, H.W., Heat
       Transfer from Flammable Fabrics, Textile Research Journal
       43, 166-175 (1973).

[35]   Miller, B. and Goswami, B.C., Effects of Constructional
       Factors on the Burning Rates of Textile Structures.
       Part I: Woven Thermoplastics, Textile Research Journal,
       Vol. 41, No. 12, December 1971.

[36]   Fourt, L. and Hollies, N.R.S., Clothing Comfort and
       Function, Marcel Dekker, Inc., New York (1970).

[37]   Renbourn, E.T. and Reese, W.H., Material and Clothing
       in Health and Disease, H. K. Lewis and Company, London (1970).

[38]   Fanger, P.O., Thermal Comfort: Analysis and Applications
       in Environmental Engineering, Danish Technical Press,
       Copenhagen (1970).

[39]   Hardy, J.D., Wolff, H.G. and Goodell, H., Pain Sensations
       and Reactions, Williams and Wilkins, Baltimore, Maryland
       (1952).

[40]   Stoll, A.M., Heat Transfer in Biotechnology--The Role of
       the Skin in Heat Transfer, Vol. 4, pp. 115-141 (ed. by
       Harnett, J.P. and Irvine, T.F.), Academic Press, New
       York (1967).

[41] Moritz, A.R. and Henriques, F.C., Studies of Thermal
     Injury, II. The Relative Importance of Time and Surface
     Temperature in the Causation of Cutaneous Burns, Am. J.
     Path. 23, 695 (1947).

[42] Moritz, A.R., Studies of Thermal Injury, III. The
     Pathology and Pathogenesis of Cutaneous Burns, An
     Experimental Study, Am. J. Path. 23, 915 (1947).

[43] Moritz, A.R., Henriques, F.C., Dutra, F.R. and Weisinger,
     Studies of Thermal Injury, IV. An Exploration of
     Causality Producing Attributes of Conflagrations; Local
     and Systemic Effects of General Cutaneous Exposure to
     Excessive Circumambient (Air) and Circumradiant Heat of
     Varying Duration and Intensity, Arch. Path. 43, 466 (1947).

[44] Henriques, F.C. and Moritz, A.R., Studies of Thermal
     Injury, I. Conduction of Heat to and through Skin and
     the Temperature Attained Therein. A Theoretical and
     Experimental Investigation, Am. J. Path. 23, 531 (1947).

[45] Henriques, F.C. and Moritz, A.R., Studies of Thermal
     Injury, V. The Predictability and Significance of
     Thermally-Induced Rate Processes Leading to Irreversible
     Epidermal Injury, Arch. Path. 43, 489 (1947).

[46] Stoll, A.M. and Chianta, M.A., Heat Transfer through
     Fabrics, Naval Air Development Center, Johnsville,
     NADC-MR-7017 (September 1970).

[47] Stoll, A.M. and Greene, L.C., Relationship between Pain
     and Tissue Damage due to Thermal Radiation, J. of
     Applied Physiology 14, 373-382 (1959).

[48] Arnold, G., Fisher, A.L. and Frohnsdorff, G., The Inter-

action Between Burning Fabrics and Skin, Unpublished Report
to Cotton Incorporated (1973).

[49] Stoll, A. M. and Chianta, M. A., Burn Protection and Pre-
vention in Convection and Radiant Heat Transfer, Aerospace
Medicine, 39 #10, 1097-1100 (October 1968).

[50] Webster, C. T., Wright, H. G. H. and Palmers, H., Heat
Tranfer from Burning Fabrics, Journal of the Textile
Institute, 53, T29-T37 (1962).

[51] Alvarez, N. J. and Blackshear, P. L., Burning Rate and Heat
Transfer from Vertically Burning Cotton Cloth Panels, Journal of Fire and Flammability, 1, p. 48 (1970).

[52] Brown, W. R. and Vassallo, F. A., Fabric Flammability Test
Development, Cornell Aeronautical Laboratory, Buffalo,
New York, CAL Report VH 2856-Z1 (May 1970).

[53] Robertson, A. F., Exploratory Studies of Heat Transfer
and Burning Behavior of Films and Fabrics, National
Bureau of Standards, Report 10-397 (1972).

[54] Burke, J., M.D., Personal Communication, Feburary 1974.

LIST OF FIGURES

1-1	Conceptualized Trade-off Analysis	8
1-2	The Decision Tree	9
2-1	Single-Layer Materials Built-up from Sub-Layers of Different Structures	25
2-2	Schematic Representation of the Test Materials Consisting of 1-4 Material Layers	27
2-3	Thermal Conductance as Functions of Temperature	38
2-4	Temperature Dependence of Shrinkage of Polyester Yarns (Diolen 100 dtex 36 Filaments) under Various Pre-stress Conditions	52
2-5	Shrinkage Stress as a Function of Temperature Measured for Polyester Yarns (Diolen 100 dtex 36 Filaments) at Given Prior Elongations	53
3-1	Energy Balance Over a Thermally Thin Slab	113
3-2	Semi-Infinite Slab	113
3-3	Heating Response of Inert Slabs	120
3-4	Slab of Finite Thickness, δ	121
3-5	Heating Response of an Inert Slab of Finite Thickness, δ	123
3-6	Fabric-Skin (Simulant) System	126
3-7	Fabric Back-Surface Temperature and Temperature Rise Near Surface of Skin Simulant vs. Time: For 100% Cotton Fabric #5, 1/4" Fabric-Skin Gap, 12.8 BTU/ft^2 $\cdot$ s	128
3-8	Fabric Back-Surface Temperature and Temperature Rise Near Surface of Simulant vs. Time: For 100% Acetate Fabric #13, 1/4" Fabric-Skin Gap, 12.8 BTU/ft^2	129
3-9	Arrangement of Heater, Shutter, Sample, and Sensors in Ignition Time Apparatus	132
3-10	Ignition and Melting Response of GIRCFF Fabrics from Wulff	134
3-11	Ignition and Melting Response of GIRCFF Fabrics	135
3-12	Comparison of Ignition and Melting Response for Individual Fabrics	140
3-13	Generalized Correlation of Polymer Ignition Data Calculated from a Mathematical Model	144
3-14	Correlation of Theoretical (A) and Semi-Empirical (D) Fabric Ranking Parameters	147

3-15   Fabric Ranking Parameter (A) vs. Mass/Unit
        Area (ρδ)   148

3-16   Fabric Ranking Parameter (D) vs. Mass/Unit
        Area (ρδ)   149

3-17   Schematic of Ignition Apparatus   153

3-18   Top and Bottom Views of Convective Ignition
        Time Apparatus   154

3-19   Schematic of Convective Ignition Time Apparatus
        with Air, Fuel, and Water Flow Paths   155

3-20   Schematic of Convective Ignition Time Apparatus
        with Transport System   156

3-21   (A) Schematic of Shutter Activation Electrical
        Circuit, (B) Schematic of Fabric Destruction
        Detection and Destruction Time Measuring
        Instrumentation   158

3-22   A Comparison of FMRC's and GIT's Experimental
        Methods   159

3-23   Projection of Specimen Loop on Plane of Burner
        Scale 0.5" for 1"   160

3-24   Flame-Spreading Speed vs. Mass/Unit Area for
        Cellulosic Materials of Different Air
        Permeabilities   163

3-25   Normalized Destruction Time vs. Normalized
        Convective Heat Flux   166

3-26   Comparison of FMRC's and GIT's Ignition Times
        for Pre-Mixed Air-Methane Flames; Fabric #10,
        Placed in a Horizontal Plane   167

3-27   Effect of Fabric Weight for Fabrics Containing
        Cellulose   169

3-28   Effect of Height above Stove for 3.80 oz/yd^2
        Cotton Fabric   171

3-29   FMRC's Minimum Response Time vs. Fabric Ranking
        Parameter, E   175

3-30   Comparison of GIT's Radiative and Convective
        Results for Cotton Fabrics   177

3-31   Comparison of GIT's Radiative and Convective
        Results for 65/35 Polyester/Cotton Fabrics   178

3-32   Comparison of GIT's Radiative and Convective
        Melting Results for Synthetic Fabrics   179

3-33   Correlation of Fabric Ranking Parameters
        Characteristic of Radiative (A) and Flaming
        (E) Ignition   181

4-1   Description of a Burning Vertically-Suspended
        Specimen   191

4-2   Relation between Front Distances and Time as
        Shown by Ignition-Time and Burning-Time
        Diagrams   192

4-3   Length of the Destroyed Specimen as a Function
        of the Igniting-Flame Time   193

4-4   Length of the Destroyed Specimen as a Function
        of the Igniting-Flame Time and Burning Time   194

4-5   Effect of Fabric Weight on Average Flame
        Propagation Velocity, Horizontal Burning   202

4-6   Effect of Fabric Weight on Average Flame Speed
        (45° Up-Burning, 1/2" Spacing, Skin below Fabric
        65% R.H. at 70°F)   204

4-7   Effect of Fabric Weight on Average Flame Speed
        (90° Up-Burning, 1/2" Spacing) (Numbers
        Represent GIRCFF Fabric Code)   205

4-8   Effect of Spacing on Burn Injury and Average
        Flame Speed (90° Up-Burning, #10 Fabric)   207

4-9   Effect of Spacing on Burn Injury and Average
        Flame Speed (90° Up-Burning, #18 Fabric)   208

4-10   Examples of Flame Spread Patterns on Dresses
        Ignited at the Front Hem   215

4-11   Replicate Garment Burns on Mannequin   216

4-12   Area of Torso of 6X Mannequin Burned by a Single
        Layer of Fabric (Dress), as related to Fabric
        Weight, Fiber Content, and Time   218

4-13   Duration of Burn for Cotton and Polyester/Cotton
        Dresses Burned on the 6X Mannequin   219

4-14   Area of Torso of 6X Mannequin Calculated to
        have Second-Degree Burns at End of Fire   221

4-15   Area of Torso of 6X Mannequin Burned by Dress-
        Slip Combinations as Compared to Dresses Along
        (Data from Fig. 4-12)   223

4-16   Effects of (A) Belt and (B) Interchange of Dress
        and Slip Fabrics on the Area of the Torso of the
        6X Mannequin Burned by Dress and Slip Combina-
        tions   224

4-17   Flame-Spreading Speed vs. Mass/Unit Area for
        Cellulosic Materials of Different Air
        Permeabilities   228

4-18   Rate of Vertically Upward Flame Propagation as
        a Function of the Weight per Unit Area   229

4-19   Vertically Upward Flame-Spreading Speed vs.
        Mass/Unit Area for Materials of Different
        Construction, via [31]   231

5-1   Section of Human Skin, after CIBA   239

5-2   Tissue Damage Rates vs. Temperatures   245

5-3   Temperature-Time Histories during and
         following Thermal Irradiation at Two Levels
         of Intensity   247

5-4   Damage during Heating and Cooling   248

5-5   Comparison of % Damage during Heating and
         Cooling   250

5-6   Comparison of Strength-Duration Curves for White
         Burns Produced by Flame Contact and by Thermal
         Radiation   251

5-7   Human Skin Tolerance Time vs. Absorbed Thermal
         Energy Delivered in a Rectangular Heat Pulse   252

5-8   Human Skin Tolerance Time Indicated by the
         Temperature Rise Measured in a Skin Simulant at
         3 sec Exposure to a Rectangular Heat Pulse   253

5-9a   Burn Depth of White Rat Skin as a Function of
         Total Heat Input from Fabrics Burning Vertically
         Upwards at 1.5" from the Skin   256

5-9b   Depth of Skin Damage (as Determined by MIT)
         as a Function of Total Heat Input   260

5-10   Average Heat Transferred to Burned Area by
         Dresses Alone   276

5-11   Heat Input Distribution Curves for Cotton Dresses
         Burned on the 6X Mannequin   280

5-12   Rate and Extent of Burn Injury in Mannequin
         Studies   282

5-13   Average Heat Transferred to Burned Areas by
         Dress-Slip Combinations   285

5-14   Effect of Fabric Weight on Burn Injury,
         Horizontal Burning, 1/2" Spacing   295

5-15   Effect of Fabric Weight on Depth of Damage
         (45° Up-Burning, 1/2" Spacing, Skin below
         Fabric, 65% R.H. at 70°F)   300

5-16   Effect of Fabric Weight on Burning Injury (90°
         Up-Burning, 1/2" Spacing)   301

5-17   Total Surface Heat Flux and Temperature
         Distribution in the Air Gap, Fabric #10(B),
         Horizontal, 1/2" Spacing   310

5-18   Temperature Distribution in Fabric-Skin Air Gap
         at Various Times during Flame Spread, Fabric
         #10(B) Horizontal, 1/2" Spacing   311

5-19   Total Surface Heat Flux and Temperature
         Distribution in the Air Gap, Fabric #10(B),
         90° Up-Burning, 1/2" Spacing   315

5-20   Total Surface Heat Flux and Temperature
         Distribution in the Air Gap, Fabric #18, 90°
         Up-Burning, 3/8" Spacing   322

5-21   Total Surface Heat Flux and Temperature Dis-
        tribution in the Air Gap, Fabric #18, 90°
        Up-Burning, 1/8" Spacing   323

LIST OF TABLES

2-1   Commercial Designations of Fabrics used in
       GIRCFF Flammability Research   28

2-2   Fabrics used in GIRCFF Research   30

2-3   Fabric Identifications and Specific Mass   33

2-4   Specific Heat of Fabrics in Ws/(gC) as
       Function of Temperature   36

2-5   Thermal Conductance Secondary GIRCFF Fabrics
       at Contact Pressure of 866 N/M^2 39

2-6   Thermal Conductance of Primary GIRCFF Fabrics   41

2-7   Auto and Pilot Ignition and Melting Temperature   43

2-8   Optical Properties of Original Fabrics   46

2-9   Optical Properties of Charred Fabrics   48

2-10   Heat Release Values of GIRCFF Fabrics   49

3-1   Experimental Results   127

3-2   Charred Fabric Optical Properties used in
       Generating Fig. 3-11   138

3-3   Fabric Ranking with Respect to Lower Bound on
       Ignition Times under Radiative Heating   150

3-4   Fabric Ranking with Respect to Lower Bound on
       Melting Times under Radiative Heating   150

3-5   Minimum Ignition and Melting Times for GIRCFF
       Fabrics   164

3-6   Probability of Ignition at 5, 10, or 15 sec
       Exposure   170

3-7   Fabric Ranking with Respect to Lower Bound on
       Ignition Times under Convective Heating   173

3-8   Fabric Ranking with Respect to Lower Bound on
       Melting Times under Convective Heating   173

3-9   Comparison Radiative and Convective
       Ignition Parameter   183

4-1   Effect of Fabric-Skin Spacing, Horizontal
       Burning   200

4-2   Effect of Fabric-Skin Orientation on Flame
       Speed, 1/2" Spacing   210

4-3   45° Up-Burning of Wool and Pure Synthetics as
       Composites with Fabric #5 as Undercloth (100%
       Cotton T-Shirt), 1/2" Spacing (65% Relative
       Humidity at 70°F)   212

5-1   Thermal Sensations and Associated Effects
       throughout Range of Temperatures Compatible

	with Tissue Life	240
5-2	Thermal Properties and Optical Properties of Skin	242
5-3	Heat Transfer to Sensing Board	268
5-4	Heat Generation Capacity of Textiles	270
5-5	Heats Released by Burning GIRCFF Fabrics under Normal Atmospheric Conditions	278
5-6	Effect of Fabric-Skin Orientation on Thermal Injury, 1/2" Spacing	293
5-7	45° Up-Burning of 100% Cotton Fabrics, 1/2" Spacing	298
5-8	Upward 45° Burning of Wool and Thermoplastics with a Cotton Knit Fabric Backing	303
5-9	Down-Burning at 45° of Thermoplastic Fabrics over a Knit Cotton Fabric	307
5-10	Total, Conductive and Radiative Heat Fluxes to Skin during Horizontal Burning of Fabric #10 at 1/2" Spacing	313
5-11	Thermal Dosage to Skin in Horizontal Burning at 1/2" Spacing (Repeat Tests)	314
5-12	Comparison of Total Conductive and Radiative Heat Fluxes to Skin	316
5-13	Comparison of Heat Fluxes to Skin in Burn Test of Fabric #18, (90° Upward at 1/2" Spacing)	318
5-14	Comparison between Heat Transfer and Thermal Injury	320
5-15a	Summary of Heat of Combustion, Heat of Burning, and Heat Transferred to Skin under Various Conditions	327
5-15b	Summary of Heat of Combustion, Heat of Burning, and Heat Transferred to Skin under Various Conditions	328
5-16	Estimate of Heat Transferred to Skin from MIT's Depth of Burn Measurements and GRI's Pathological Studies in Rats	329
5-17	Comparison of Burning and Heat Transfer Efficiencies	334

Absorptance (optical), 46, 48
Acetate fabrics, 30
 thermophysical properties, 85,
 91
Acrylic fabrics, 30
 thermophysical properties, 77
Air permeability, 57-65, 227-230
 GIRCFF fabrics, 30, 66-106

Burn injury, 236-337, 351-335
 degree of burn, 240, 352
 from fabric combinations, 299
 from fabric strips, 265, 287
 effects of orientation on, 290
 294, 297
 effects of spacing on, 290,
 294, 297
 skin simulants, 288
Burn injury criteria, 241, 260
 energy density, 254
 Henrique's, Stoll's, 241
 MIT's, Henrique's, 258
Burn injury from garment,
 273-287
 fabric combination, 283
 local heat transfer, 279
 overall heat transfer, 271, 273

Charred fabrics, 66-106
 optical properties, 48, 138
Cotton fabrics, 30
 thermophysical properties, 73,
 74, 80, 83, 102, 104
Cotton/polyester
 see Polyester/cotton

Double knit fabrics, 71
Durable press fabric, 66, 96

Ease of extinguishment
 garment fires, 226

Fabrics
 see also Structures,
 textile fabrics
 burning residue, 49
 compositions and combinations
 in real-life situations, 27
 GIRCFF series, 30
 thermophysical properties,
 31-56, 66-106

Fabric structure
 see Structures, textile
 fabrics
Fire retardant fabric, 104
Flame spreading in fabric
 strips, 195-213
 comparison of strips vs.
 garments, 226
 effect of air permeability on,
 227-230
 effect of fiber composition,
 196-199
 effect of orientation on, 209
 effect of spacing on, 199
 effect of structure on, 227-232
 in fabrics composites, 209
Flame spreading in garments,
 213-227
 comparison of single and combinations, 222
 comparison of strips vs. garments, 226
 ease of extinguishment, 226
 garment combinations and
 design, 220
 single garments, 213

Heat of burning, fabrics, 47-51
Heat transfer analyses, 111-124,
 139-143
 thermally thin slab, 112-116
 thermally thick slab, 116-124
Heat transfer to skin, 306-337
 conduction and radiation,
 308-309
 effects of fabric-skin spacing,
 319-324
 from dripping molten polymer,
 267-269
 from horizontal samples, 309-314
 from vertical samples, 314-324

Ignition, 107-182, 342-345
 apparatus, 131, 152-160
 fabrics, 109
 fabric sleeves as by ranges,
 157, 168
 flaming, 152
 heat transfer analysis, 111-124
 139-143
 radiative, 131

Ignition (continued)
  radiative vs. flaming, 174
  ranking of ease of ignition
    or melting, 137, 150, 168,
    173
  versus melting, 110
Ignition temperature, 40, 43
  GIRCFF fabrics, 66-106

Knit fabrics, 30, 74, 77, 79

Layering of materials, 27
Limiting oxygen index, 66-106

Mass/unit area, 33, 66-106
Melting temperature, 40,43
  66-106
  versus ignition, 110
Micrographs, fabrics, 67-105

Nylon fabrics, 30, 85, 87, 93

Optical properties, 46, 48,
  66-106
Optical thickness, 46, 48,
  66-106

Pre-ignition heat transfer,
  124-130
Polyester/cotton fabrics, 30,
  66, 76, 79, 98, 99
Polyester fabrics, 30, 68, 71,
  96

Radiant properties of fabric,
  44-47
Radiant properties of skin, 242
Ranking of ease of ignition or
  melting, 137, 168
  flaming heat sources, 168, 173
  radiant heat source, 137, 150
Reflectance (optical), 46, 48
Residue of burning fabrics, 49

Scanning electron micrographs,
  67-105
Shrinkage stress, thermal 51-56
Shrinkage, thermoplastics, 52
Skin description, 238-241
  pre-ignition heat transfer to,
  124-130
  simultants, skin, 288-290

thermal and optical properties of, 242
Specific heat, GIRCFF fabrics,
  32-36
Structures, textile fabrics,
  23-31
  see also Double knit fabrics
  see also Knit fabrics
  see also Warp knit fabrics
  see also Terry cloth
  commercial designation, 28
  effect of flame spreading,
    227-232
  effect of temperature, 51-56
  scanning electron micrographs,
    57-65
Terry cloth, 80
Textile fabric structures
  see Structures, textile fabrics
Thermal conductance, 38, 39, 41
Thermophysical properties, GIRCFF
  fabrics, 31-65, 66-106
  absorptance, 46, 48
  ignition temperature,43
  mass/unit area, 33
  melting temperature, 43
  optical properties, 46, 48
  optical thickness, 46, 48
  reflectance, 46, 48
  specific heat, 36
  thermal conductance 38, 39, 41
  transmittance, 46, 48
Thermophysical properties, skin,
  242
Thermoplastic textiles
  acetate fabrics, 30, 85, 91
  acrylic fabrics, 30, 77
  ignition vs. melting, 110, 111
  nylon fabrics, 30, 85, 87, 93
  polyester fabrics, 30, 52, 53,
    68, 71
  shrinkage, 52
  shrinkage stress, 53
Transmittance (optical), 46, 48,
  66-106

Warp knit fabrics, 30, 85, 87, 91
Wool fabrics, 30, 103, 106